In All Probability

Investigations in Probability and Data Analysis

GEMS® Teacher's Guide for Grades 3–5

by
Jaine Kopp

Skills

Describing • Comparing • Predicting • Estimating
Collecting and Recording Data • Organizing Data • Analyzing Data
Making Inferences • Conducting Probability Experiments
Applying Probability Concepts • Practicing Computation
Using Classroom Discourse (Speaking, Listening, Writing, Reading)

Concepts

Fair and Unfair Chances • Equally Likely Events
Experimental Probability • Theoretical Probability
Law of Large Numbers • Graphing • Data Interpretation
Rational Numbers • Combinatorics (Combinations and Permutations)

Mathematics Strands

Data Analysis • Probability • Number and Operations

Themes

Patterns of Change • Models and Simulations

Time

13 or more 45-60 minute sessions

GEMS gratefully acknowledges the generous financial support of The Moody's Foundation for the field testing, revision, and production of this New GEMS version of *In All Probability.*

Great Explorations in Math and Science
Lawrence Hall of Science
University of California at Berkeley

Lawrence Hall of Science,
University of California,
Berkeley, CA 94720-5200

Original 1993 edition authored by
Celia Cuomo

Initial support for the origination and publication of the GEMS series was provided by the A.W. Mellon Foundation and the Carnegie Corporation of New York. Under a grant from the National Science Foundation, GEMS Leaders Workshops were held across the United States. GEMS has also received support from: the Employees Community Fund of Boeing California and the Boeing Corporation; the people at Chevron USA; the Crail-Johnson Foundation; the Hewlett Packard Company; the William K. Holt Foundation; Join Hands, the Health and Safety Educational Alliance; the McConnell Foundation; the McDonnell-Douglas Foundation and the McDonnell-Douglas Employee's Community Fund; the Microscopy Society of America (MSA); the NASA Office of Space Science Sun-Earth Connection Education Forum; the Shell Oil Company Foundation; the University of California Office of the President; and the Moody Foundation. GEMS also gratefully acknowledges the early contribution of word-processing equipment from Apple Computer, Inc. This support does not imply responsibility for statements or views expressed in publications of the GEMS program. For further information on GEMS leadership opportunities, or to receive a publications catalog please contact GEMS. We welcome letters to the *e-GEMS Network News* and encourage you to sign up to receive this occasional newsletter at www.lhsgems.org

ISBN-13: 978-1-931542-07-4
ISBN-10: 1-931542-07-4

Printed on recycled paper with soy-based inks.

Library of Congress Cataloging-in-Publication Data

Kopp, Jaine.
In all probability : investigations in probability and data analysis : GEMS teacher's guide for grades 3/5 / by Jaine Kopp.
p. cm.
Includes bibliographical references.
ISBN-13: 978-1-931542-07-4 (pbk.)
ISBN-10: 1-931542-07-4 (pbk.)
1. Probabilities--Study and teaching (Elementary)--Guidebooks. 2. Statistics--Study and teaching (Elementary)--Guidebooks. I. GEMS (Project) II. Title.
QA273.2.K67 2006
372.7--dc22

2006009775

Acknowledgments

- Celia Cuomo, friend and former Math Education Program colleague, who authored the original GEMS version of *In All Probability* and sparked us all to think more about how to teach data analysis and probability to elementary students;

- Kimberly H. Seashore, my Bay Area Mathematics Project colleague and comrade, who helped think through the activities with an eye on the mathematics content;

- Dr. Martin Weissman, mathematician at UC Berkeley, who reviewed the guide for mathematical accuracy;

- Elaine Ratner, who provided crucial and timely editorial assistance;

- Francesca DeLuca and her fourth grade class at St. Jarlath School in Oakland for participating enthusiastically in the revised activities and providing encouraging and helpful feedback;

- The California teachers who trial-tested this 2006 version of *In All Probability* and provided valuable feedback are listed on this page. teachers who field-tested the 1993 edition are listed in the back of the guide;

- The children from St. Jarlath and Park Day schools in Oakland, who generously allowed us to photograph them in action;

- Alan Goodfried, my husband, for taking the photographs at St. Jarlath School and Richard Hoyt and Laurence E. Bradley for the earlier photographs, all of which capture young people actively and enjoyably engaged in mathematical learning.

2006 Trial Test Teachers

Bakersfield

Michael Cushine
Van Horn School
Janice Karnowski
Columbia Elementary School
Terie Lynne Storar
Valle Verde Elementary School

San Francisco

Susan DesBaillets
Grattan Elementary School

San Pablo

Antoineta Franco
Emily Vogler
Downer Elementary School

Contents

Time Frame

Based on classroom testing, the following are guidelines to help give you a sense of how long the activities may take. The sessions may take less or more time with your class, depending on students' prior knowledge, their skills and abilities, the length of your class periods, your teaching style, and other factors. Try to build flexibility into your schedule so that you can extend the number of class sessions if necessary.

Activity 1: Penny Investigations

Session 1: Introducing Data .... 45–60 minutes
Session 2: Penny Flip .... 45–60 minutes
Session 3: Making "Cents" (Sense) of the Penny Data ..... 45–60 minutes
Session 4: Graphing Heads and Tails .... 45–60 minutes
Session 5: Introducing Theoretical Probability .... 45–60 minutes

Activity 2: Spin

Session 1: Ready, Set, Spin! .... 45–60 minutes
Session 2: What's the Real Spin on the Spinners? .... 45–60 minutes
Session 3: Spinner Investigation: Fair or Not Fair? .... 45–60 minutes

Activity 3: Horse Racing

Session 1: Roll of a Die .... 45–60 minutes
 Introduction to a Die .... 10 minutes
 Experiment 1: Horse Race .... 30 minutes
 Experiment 2: Roll ALL Six .... 45 minutes
Session 2: Off to the Races! .... 30–45 minutes
Session 3: Keeping Track .... 45–60 minutes

Activity 4: Game Sticks

Session 1: Game Sticks in Action .... 45 minutes
Session 2: Investigating the Outcomes .... 45–60 minutes

INTRODUCTION

This unit combines essential standards-based mathematics content and inquiry opportunities with real-world relevance and highly motivating experiences.

Your students experience probability and statistics frequently in their daily lives. They play games with spinners, coins, cards, or dice, and all of these involve probability. They encounter data on food packaging, on television commercials, in news stories, on the sports page, on baseball cards, in political opinion polls, election results, or on graphs of the school fund-raising drives. This unit taps into your students' real-world experiences and engages them in exciting investigations that develop standards-based mathematics literacy in data analysis and probability. The crux of the mathematics is data collection, organization, representation, and interpretation to understand outcomes of probability experiments.

Before you begin, it's highly recommended that you read through the **Background for Teachers** section in the back of this guide to refresh your content knowledge in this realm of mathematics. This background will assist you in facilitating classroom discourse and building conceptual understanding. Within the compelling and sometimes surprising context of conducting probability investigations, students generate data for themselves, graph, interpret data, and communicate reasoning and results. They write about what they are learning and talk about it to each other and with you. They display, represent, and convey information, findings, and experimental results. They predict, describe, compare, compute, and draw conclusions. They apply logical thinking skills and gain insight into how the major mathematical ideas of probability and data analysis relate to them and their world. These key concepts and thinking skills will also serve them well in other areas of learning, such as science, social studies, and the language arts.

Helping students understand how often and in how many ways their lives and decisions relate to probability and statistics can be one of the most valuable and lasting outcomes of this unit. Such knowledge will be very important to them as citizens and decision-makers, whether it be in making an informed consumer choice, evaluating economic statistics in newspaper articles, judging the chances of winning a lottery, or any of hundreds of other moments and decisions in daily life.

This New GEMS™ transformation of the classic GEMS guide has four main activities. The activities are still built around the same types of probability experiments (penny flip, spinners, dice, Native American game sticks) and reflect the GEMS step-by-step presentation format—but a great deal has changed. Among the most important changes are increased emphases on flexible approaches to gathering data and to data analysis, and on the use of mathematical discourse to build individual and class content knowledge. These new emphases in turn led to important shifts in the structure of the unit. These changes are in line with current 21st century thinking in mathematics education and cognitive research, and with state and national standards, including recommendations of the National Council of Teachers of Mathematics (NCTM).

Into the Guide

The four main activities in the new unit are described below. More detailed descriptions of each class session are provided in the Overview to each of the main activities.

Activity 1: Penny Investigations has five class sessions. It sets the stage and provides students with foundational tools for the entire unit. Pennies provide the springboard into the investigations and the context for collecting, organizing, discussing, and interpreting data. Students conduct the Penny Flip experiment to generate and record data, and to interpret data results. The concept of fairness and vocabulary related to probability are introduced and defined. Scales for graphing are discussed and students critically examine graphs from newspapers and advertisements. Embedded in the activities are opportunities for computational practice.

Activity 2: Spin to Win consists of three class sessions involving probability experiments with "fair" and "unfair" spinners to make the moves in two Track Meet races. Students gather data, analyze the data, and make conjectures about fairness based on the data. The spinners also provide a context for understanding fractional parts of a whole. Student understanding is deepened through classroom discourse and questioning strategies to probe and encourage student explanations of reasoning.

Activity 3: Horse Racing has three class sessions involving dice. Teachers choose the first experiment—either the Horse Race or the Roll ALL Six game. With either, students roll one standard die and analyze the outcomes for many rolls. Then the students play the Double Dice Derby game with two dice and keep track of the winners of the races on a class graph. At home, they race the horses to gather more data and report the wins on the class graph. The class graph is also represented as a line graph to provide a different way of looking at data. Using a Keeping Track chart, students record all possible sums for two dice in an organized way to better understand the results of the race. The theoretical probability for the outcome of two dice is also considered.

Activity 4: Game Sticks introduces students to a version of a Native American game of chance. After creating their own set of sticks with colorful designs, students use the sticks to play the lively game. Then students conduct an experiment to analyze the outcomes (combinations of plain and design sides) of the sticks. The class pools their results to get a better picture of the frequency of each outcome. Finally, they use the sticks to determine all the variations (permutations) for each outcome. The variations are connected to the theoretical probability of their occurrence.

The guide includes **Summary Outlines** for all activities for quick reference and to assist you in guiding your students through these activities in an organized fashion. We recommend that you go through the activities as detailed in the guide, activity by activity, the first time you teach the unit. After you are familiar with the development of content, the Summary Outlines may be helpful to streamline preparation and serve as handy presentation notes.

Standards Connections to *In All Probability*

The mathematics activities included in this guide are rich with opportunity for students in Grades 3 through 5 and have been used successfully, with appropriate modifications, with both younger and older students. You can decide how to present the activities based on the ability and experience of your students. The activities address the National Council of Teachers of Mathematics (NCTM) national standards, outlined in *Principles and Standards for School Mathematics*, as well as state standards and frameworks. The content in the Data Analysis and Probability and Number and Operations standards is developed through the course of the unit, and the process standards—Problem Solving, Reasoning and Proof, Communication, Connections, and Representation—are embedded in the presentation of each activity.

The following Standard lists the core mathematics content for this unit:

NCTM Standards for Data Analysis and Probability
For Grades 3-5 (listed for the span to be achieved by end of Grade 5)

Formulate questions that can be addressed with data and collect, organize and display relevant data to answer them.

- Design investigations to address a question and consider how data collection methods affect the nature of the data set.
- Collect data using observations, surveys and experiments.
- Represent the data using tables and graphs (plots, bar graphs, line graphs).
- Recognize the difference in representing categorical and numerical data.

Select and use appropriate statistical methods to analyze data.

- Describe the shape and important features of a set of data and compare related data sets, with an emphasis on how the data are distributed.
- Compare different representations of the same data and evaluate how well each representation shows important aspects of the data.

Develop and evaluate inferences and predictions that are based on data.

- Propose and justify conclusions and predictions that are based on data and design studies to further investigate conclusions or predictions.

Understand and apply basic concepts of probability.

- Describe events as likely or unlikely and discuss the degree of likelihood using such words as certain, equally likely and impossible.
- Predict the probability of outcomes of simple experiments and test the predictions.
- Understand that the measure of the likelihood of an event can be represented by a number from 0 to 1.

All activities involve your students in conducting probability experiments to generate data. With guidance from the teacher, students create their own data sheets to record the results of the experiments, and they organize the data in the form of charts and graphs to interpret it. The data analysis provides opportunities for critical thinking and making conjectures based on fact. This process develops students' ability to design the tools needed to gather and display data that will be used throughout their mathematics education. See "Data Representation," starting on page 106 for more detailed information. Using the data, students determine the "fairness" of a game or the likelihood of an event and represent this on a scale from 0 to 1.

Teachers have reported positive changes in their students' abilities to generate, record, and interpret data after they have had the challenge of making their own data sheets and graphs in this unit.

Academic Literacy Development

Mathematics provides a rich context for students to develop literacy skills and academic language. Specific mathematics language is developed through the content, and distinctions are made between common usage of English words and their mathematical meaning. The language for probability and data analysis is built into and practiced throughout the activities. Students engage in verbal discourse throughout the unit and develop the ability to make a case for their thinking. Listening skills are developed during class discussion and debate. Writing is used as another vehicle for students to reflect on the events or discoveries of the lesson, recast their learning into their own words, and clarify their thoughts as they transfer their ideas to paper. Some teachers use questions related to the mathematics students are studying as "starters" during writing time. Several writing suggestions are included with each activity, and you are encouraged to adapt these prompts or create you own.

Assessment

Throughout the guide, there are opportunities to assess students' understanding. See page 132 for a summary of specific assessments within each activity as well as additional assessment suggestions. There is also an End of Unit Assessment to measure growth and determine student proficiency with content, and to help identify misconceptions or gaps in knowledge.

Closing Note

As you present the activities in this guide, look for evidence (which we "predict" you'll find!) that students are interested and curious, make predictions, ask questions, interpret data, and suggest other probability experiments. The chances are great that these increasingly complex opportunities for students to experiment mathematically will open up a whole new world of probability to them! By asking thought-provoking, open-ended questions—and stimulating rich classroom discussion and discourse—you are setting up an environment where learning can, and in all probability will, flourish! ■

Probability Poetry

by L.B.

Prediction's the art of horse before cart
A race against time, a flip of the dime
Most probably yes, quite possibly no,
Perhaps it will happen, maybe not so,
Things that we think most likely to be
May or may not come to pass you see,
The chance of a snowball that falls toward the Sun
Or the order in which a race will be won
The roll of a die or how game sticks will fall,
Probability helps us to look at it all—
To estimate what we think will take place
In a toss of dice for a great horse race
To experiment, more data to gain,
Then make a graph to help explain
The reasons why when two dice fall
Some numbers come often, one not at all.
Our predictions then aren't guesses wild
They're based on evidence we've compiled.
A crystal ball could cloud over my friend
But on mathematics you can depend—
Probability works, that's undeniable,
But not every time, it's not that reliable
It helps us predict how things may come out
And that's enough to be happy about!
The law of large numbers we learned, for example,
That odds come closer as sample's more ample!
These lessons and games give you practice and fun
And you'll be an expert before you are done—
Because 9 out of 10 young mathsters agree
You'll get into *In All Probability!*

Overview

This first activity sets the stage for this unit on data analysis and probability. Pennies are the springboard into the investigations and provide the context for discussing, collecting, organizing, and interpreting data. The focus turns to probability as the results from an experiment are analyzed.

In Session 1, students begin by sharing their prior knowledge about data and pennies. As students report what they know about data, their ideas are recorded. The penny is reviewed as a coin in our monetary system with a value of one cent (.01 of $1.00). Students closely examine pennies and discuss the components of the coins. Then the pennies are organized and graphed, using one piece of data on each coin—the year minted. For homework, students examine pennies at home in search of the oldest penny.

Session 2 opens with students sharing their pennies from home. The oldest penny is identified and the pennies from home are added to the class graph. Next, students estimate the number of pennies in a small jar and determine the actual number. The pennies are then used to conduct an experiment—Penny Flip. Students make data sheets to record 10 flips, predict their outcome for 10 flips, and work with a partner as they flip and record tosses. Their results are recorded on a class graph and discussed. Then the students conduct the experiment a second time and add those results to the graph. The students analyze how the graph has changes with additional data added and make conjectures about the comparative number of heads and tails.

In the next session, the data is analyzed through the lens of the total number of heads and tails tossed to get a clearer picture of the outcomes of flipping a penny. Students use computational skills to calculate the number of heads and tails. With this data, they make a graph to represent the comparison of heads to tails. The critical part of this graphing exercise is to determine an appropriate scale to record the data. For homework, students create another graph of the penny flip data to compare heads and tails on grid paper.

At the start of Session 4, students share their graphs with one another. The graphs provide a vehicle to reinforce the need for a consistent scale and a scale that reflects the amount of data collected. A graph for an advertisement is analyzed, so that students see the effect of using a scale that easily leads to misconceptions about the data. Pairs of students examine other graphs to become aware of scales and their impact on the information the graph displays.

In Session 5 the theoretical probability behind a penny flip is examined. The vocabulary and language of probability are introduced and students connect real-world situations to the numbers zero (never happen) and one (always happen). In the case of a penny flip, it is equally likely to toss a head or a tail and this is expressed as one half. A number line is labeled and used to situate the probabilities. The book *Cloudy with a Chance of Meatballs* is an optional springboard to the initial discussion of probability.

Where's the Math?
This activity provides opportunities for students to gather data related to pennies and organize the data on various graphs. As the graphs are analyzed, students make observations and conjectures about the data. The Penny Flip introduces a probability experiment that allows students to generate and record data, and to organize and interpret data results. The concepts of "never," "always," and "equally likely" are introduced and additional vocabulary related to probability is developed. As appropriate for students, the theoretical probability is discussed and represented on a number line from 0 and 1. Embedded in the activities are opportunities for computational practice. Classroom discussion and questions focused on mathematics help students to develop conceptual understanding and math reasoning skills.

Session 1: Introducing Data

■ What You Need

- ❑ chart paper, plain (approximately 26 x 38 inches)
- ❑ chart paper, square grids (approximately 26 x 38 inches)
- ❑ markers
- ❑ pennies

For each student

- ❑ 1 penny
- ❑ math journal

Optional:

- ❑ hand lens for each student or pair

■ Getting Ready

1. During this unit, students will be collecting data, learning vocabulary, writing responses to prompts, etc. It is recommended that

students have a math journal to keep this information together, and a pocket folder for math papers (graphs) is suggested.

2. Gather chart paper and markers.

3. Prepare the graph for organizing the pennies by year. Cut a sheet of grid chart paper in half lengthwise. Tape the short sides of the halves together so that the grid lines align. Label the graph starting at 1970 and continue to current year. (See page 13 illustration.)

4. Gather an assortment of pennies with years dating from the 1970s to the current year. You need at least one per student.

5. Familiarize yourself with the information on pennies in the Background for Teachers section on pages 112–113.

6. Read the Letter to Families on page 35 to decide if you will send that letter home with your students or write one of your own. Duplicate one copy of a letter about the unit for each student, so that families will be informed about this unit.

■ Data, Data, Everywhere!

1. Begin by writing the word *data* on the board. Ask the students to think about data. Ask open-ended questions to get them started, such as:

   What do you know about data?
   What have you heard about data?
   How do you think data is used?

2. After the students have had a few minutes to think independently, have them share their ideas about data; record them on plain chart paper. The following is a sample of the ideas generated by a fourth-grade class.

   Our Ideas About Data
   Information you can find on computers
   How to explain your math problems
   Something that explains something else
   Something that has information about something
   Doctors and scientists use data, like when they investigate stars or study about the body
   Data is used in graphs
   It is like a date
   Any information you can get

Data is information, and in math, it especially refers to numerical information. At this moment, highlight the ideas that are accurate and correct. Use other ideas as teaching opportunities. For example, an English learner said it was "like a date." Write the word date under the word data and point out the similarity between the two words. Explain the difference!

3. Distribute a penny to each student. Have them identify the coin and its value. Ask the students to examine the pennies closely and find out all they can about the coin. Provide hand lenses, if available.

4. Circulate and listen to students' observations.

5. Focus the class to share their discoveries. This is an opportunity for those who are often quiet to share information, such as the year their penny was minted, with the class.

Be sure your students know the meaning of the word "mint" and "minted" in the context of coins. You may want to discuss the multiple meanings of the word mint.

6. As a student shares her findings, have the rest of the class look at their pennies to locate the information on their coins. Some items that they are likely to observe include:

   English words: *In God We Trust; Liberty; United States of America; One Cent*

   Latin words: "*E Pluribus Unum*" (out of one, many—meaning we are all one people)

   Head Side (obverse):
   - Abraham Lincoln, 16th President, 1861–1865
   - Year minted
   - Letter P (Philadelphia), D (Denver) or S (San Francisco)—city where the coin was minted, under the date

   Tail Side (reverse):
   - Lincoln Memorial—using a hand lens you can see Lincoln sitting inside
   - Two stalks of wheat—on coins minted before 1959

7. Have students focus on the year their pennies were minted. Have them find this piece of data on their coins. Ask who thinks they have the newest or most recently minted penny. (One or more students are likely to have a coin minted in the year you are doing the unit.)

8. Continue with the oldest penny. Record the proposed years for the oldest penny. Order those pennies by the year they were minted. Determine the oldest penny and calculate how old it is. Point out that this penny is the oldest *in this particular sample* of pennies!

9. Ask the students for some of the other years their pennies were minted. Record those numbers on the board. Tell the students you want to use the dates the coins were minted as the data to make a graph.

10. Gather the students in a location in your classroom that will allow your class to organize the pennies from the oldest to the most recently minted on the grid chart paper.

11. As you create this concrete graph, be sure to leave a space for each of the years that there is not a coin represented. When all the pennies are graphed, ask the students for a title or label for this graph. As a group come up with an appropriate title, such as "Our Collection of Pennies Organized by Year Minted." Here is a sample of a graph created by a class.

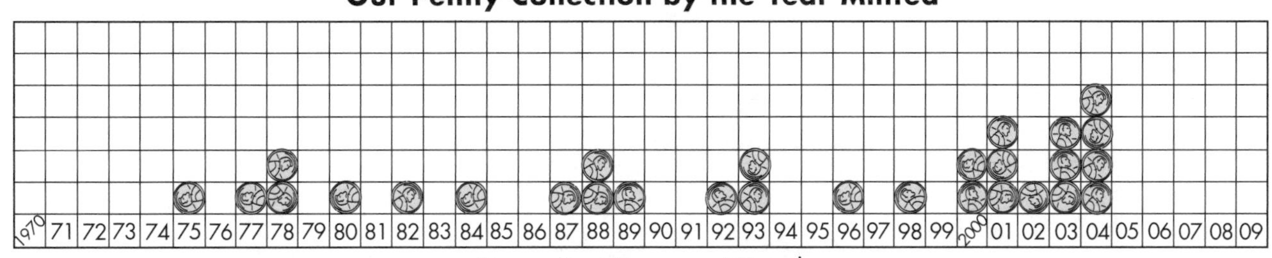

12. Identify this as a bar graph. Have students talk to a partner or in small groups about the information in the graph. Have students share their observations and interpretations. Help qualify statements with "for this collection of pennies," since this is only one small sample of coins.

13. As students share information from the graph, differentiate between the factual data and inferences about the data. For example, here are facts about the data:

On our graph, there were 3 coins minted in 2003.

The pennies in our collection were minted from 1975 to....

Versus statements such as:

Most pennies were minted after 2000.

In this case, though the graph may show that the most pennies in this collection were minted after 2000, it is not necessarily true of all pennies.

The year 2003 won.

Steer students away from viewing a graph as a competition! A graph is a representation of data. Instead, a statement for that inference would be, "In our collection, the year 2003 had the most coins."

14. At the end of the discussion, give your students a **Homework Assignment.** Have them ask their families for any pennies in their pockets and/or purses (and from coin jars!). Students are to share with their families (or at least one person) all they learned about this coin in class. Next, they should organize the pennies by the year they were minted and select the two oldest pennies from this collection. They should ask their family members if anyone knows the year the first pennies were minted, then bring the two oldest pennies to class the following day.

Session 2: Penny Flip

■ What You Need

- ❑ graph of pennies organized by year minted (made in Session 1)
- ❑ sticky notes (2 distinct colors, 2.5 by 2.5 inches square)
- ❑ 11 Index cards, 3 x 5 inches
- ❑ small clear container
- ❑ extra pennies

For each student:

- ❑ 1 penny
- ❑ math journal

■ Getting Ready

1. When a penny is tossed 10 times, there are 11 different combinations of heads and tails that can occur. Record each possible outcome on a separate index card. These cards will be used to create a graph. The outcomes are as follows:

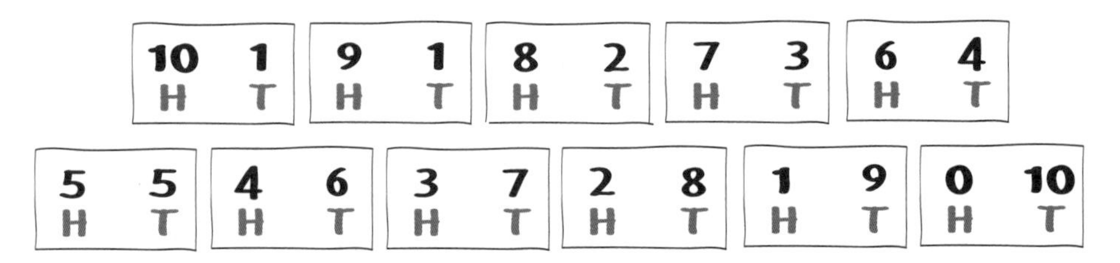

http://www.mcli.dist.maricopa.edu/mobius/flip/ This website simulates coin flips, done every five seconds. Also, the user can design and carry out coin flip experiments. It even graphs the results of the experiments in a similar way as students do in these activities.

2. Have the pennies from Session 1 available. Each pair of students will use one penny to conduct the Penny Flip experiment.

3. Fill a small jar, such as a baby food jar, with up to 59 pennies for an estimation activity.

4. Gather two pads of sticky notes in contrasting colors. The sticky notes will be used to record the flips in the penny experiment. One color will be used to record the first round of flipping and the other color the second round.

■ Pennies from Home

1. Have students recall how the pennies were organized on a graph in the opening session.

2. Segue into students sharing their pennies from home with a partner. Have them determine who has the oldest penny between them.

3. Ask the class for the dates on their oldest pennies and record them on the board. Determine which penny among the ones that are in the class is the oldest.

4. Ask if anyone found out the year when pennies were first minted. If not, tell them the first penny was minted in 1787 in Philadelphia. It was pure copper and was designed by Benjamin Franklin.

■ Estimation Jar

1. Hold up the small jar of coins. Ask students to estimate how many pennies are in the jar. Remind them that an estimate is a thoughtful guess—not a wild guess!

2. Have pairs of students talk and share their estimates. Listen to their estimates to get a range of their numbers.

3. Pour the pennies on a table or in an area where students can see them out of the jar. With this new perspective on the quantity, let them adjust their estimates.

4. Count out 10 pennies. Look at the remaining amount. Ask if the students think there are 100 pennies. Have them respond with thumbs up (yes) or down (no) or horizontal (maybe).

5. Count out another set of 10 pennies. Continue until there are 40 pennies counted. Have the students make a final estimation.

6. After the number of pennies is known, have students reflect on how their estimates changed over the course of the count.

7. Ask what other coins would equal the amount of money in the jar. Encourage the students to come up with multiple answers using quarters, dimes, and nickels and using pennies only as necessary.

Remind students that estimations of this sort are not contests! They are designed to help everyone get better at making estimations.

■ Penny Flip

1. Ask the students if they have ever flipped a coin before. Have them give reasons why people might toss coins. Ask what outcomes are possible when a penny is tossed. Do they think flipping a penny is a fair way to decide something? Why or why not?

2. Tell students they are going to use pennies to conduct an experiment. When mathematicians or scientists conduct an experiment, they often predict what might happen. Write PREDICT on the board. The word comes from Latin—PRE means *before* and DICO means *say*. Predict literally means *say before*.

3. Let them know that in this experiment they will flip a penny 10 times. Ask for a possible outcome—or the combination of heads and tails—one might get when a penny is flipped 10 times. Create a list of outcomes with the class.

4. Tell students that you are going to demonstrate a penny flip experiment. Have them predict how many heads and how many tails you will get in 10 flips.

5. Select a student partner and model how to conduct the experiment with a partner, as follows:

   a. Ask your partner to be the "flipper" for 10 flips and say you will be the "recorder" of those 10 flips.
   b. Have your partner predict how many heads and how many tails she thinks she will toss.
   c. As the recorder, on the board or overhead write an H for heads and a T for tails, as follows:

   H T

   d. Then create a grid to record tosses with tally marks and put a large square around the H and T to simulate a giant sticky notes, as follows:

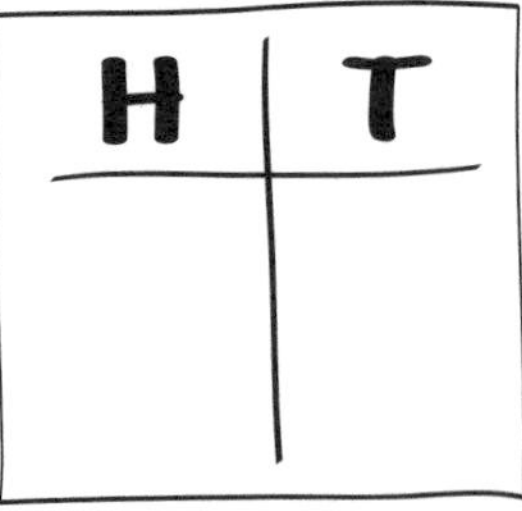

   e. Let your partner begin flipping the penny. After each flip, put a tally mark under the H or T.

   f. As the recorder, keep track of the outcomes until there are a total of 10 flips completed.

g. Count the number of heads and tails, and record the totals under the tally marks. Here is a sample of a completed set of 10 flips:

h. Switch roles so that you flip 10 times and your partner records your 10 flips.

Define a uniform way to toss or flip the penny. This helps avoid pennies flying through the air and landing all over the room. Some teachers have used felt squares on desktops to soften the sound of flipping coins.

■ Students Flip

1. After students understand the procedure, distribute one sticky note per student. Guide students in setting up the recording system on their sticky notes. Be sure they list H (heads) first and then T (tails) so that all the data will be reported in a consistent manner.

2. Distribute pennies and remind students to predict what their outcomes will be for 10 flips and record their predictions in their journals.

   As they are working, post the index cards with the pre-recorded outcomes along the bottom of the board or on a wall. Then circulate and assist students as needed.

3. Focus the group and have students record the number of heads and tails they flipped in their journals. How close were their predictions to the actual number of heads and tails? Were they surprised at any of the outcomes they got? Why?

4. Emphasize that each set of partners has generated the results of two experiments and has two pieces of data. Ask if these two pieces of data are enough to predict what is likely to happen when you toss a coin 10 times. Have students talk to their partners and listen to their ideas

5. Tell the class that to have more information to base a prediction on, they are going to combine the data generated by the whole class. Focus their attention on the outcomes you posted on the board. Tell students they are going to create a graph of the data by posting their results above the outcome that it matches.

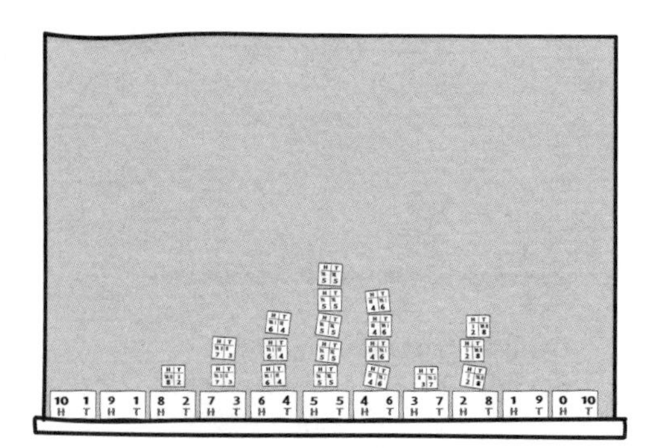

Be sure to distinguish between true statements or facts about the data and inferences. See the Background for Teachers section on pages 106–109 for more information on data interpretation.

6. Have about four students come up at a time to post the results of their flips above the combination it matches. After each group adds new data, look at the impact on the data on the graph.

7. Once all the class data is recorded, ask for observations about it. As necessary, pose questions to help guide your students, such as:

What was the most common outcome in our experiment?
Why do you think that there were so few... or so many?
Is it possible to toss 9 heads and 1 tail? Why or why not?
Is it likely you will toss 10 tails? Why or why not?
What would happen if we toss again 10 times and add the results to the graph?

■ Penny Flip Revisited

1. Tell students that they are going to gather more data on outcomes of penny flips. Partners will conduct the experiment again in the same manner. Remind them to predict the outcome before they begin.

2. Distribute a sticky note—in a contrasting color—to each student. Have them set up their recording system for heads and tails.

3. Once partners are ready, they begin. Circulate as they flip and record.

4. When the class is done, focus their attention on the graph again. Ask them to predict what will happen as more results of the penny flip are added to the graph. Have partners discuss and listen to each other's hypotheses.

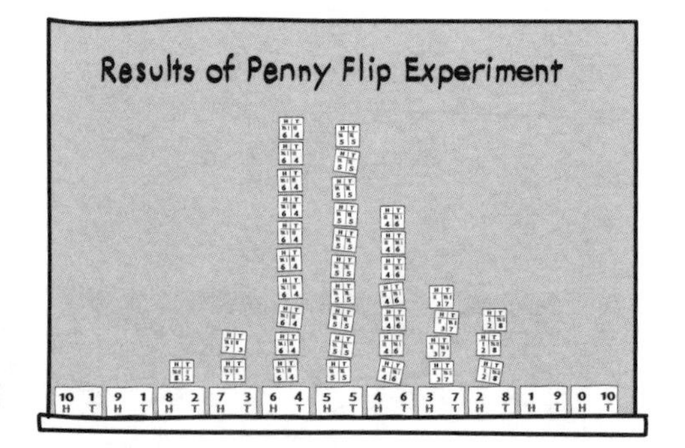

5. As before, have students record the actual number of heads and tails in their journals. Did they get the same result for their second experiment as for their first?

6. Have four students at a time post their results on the class graph. Stop at various points and have them observe how the graph changes as more data is added to it. When all the data is posted, title the graph with your students' help.

7. Let partners share their observations with one another. Then have a class discussion on what information the data provides. Be sure students explain their observations and check with the class for agreement as well as dissenting ideas.

8. In closing, have your students analyze the data in a new light. If they added up ALL the heads and ALL the tails on the graph, ask them to predict if there would be:

   More Heads than Tails (H > T)
   More Tails than Heads (T > H)
   Same Number of Heads as Tails (H = T)
   Clarify that this includes the recorded heads and tails ***from all the outcomes*** on the graph.

The data on the graph will be used in the next session.

9. Have students write their predictions in their journals and explain the reasoning behind them. Collect their journals at the end of the session to informally assess your students' understanding at this point.

Session 3: Making "Cents" (Sense) of the Penny Data

■ What You Need

- ❑ graph of the results of the Penny Flip
- ❑ centimeter (cm) grid paper (8.5 x 11 inches)
- ❑ overhead transparencies and pens

For each student:

- ❑ cm grid paper
- ❑ math journal

■ Getting Ready

1. Using the centimeter (cm) grid paper master on page 34,
   a. Make three overhead transparencies of the grid.
   b. Make one paper copy for each student for homework.

GO ■ Calculating the Number of Heads and Tails

1. Return the students' journals and have them review what they wrote about the relative number of heads and tails flipped by the class. Have students share their hypotheses with a classmate.

2. Tell students that they are going to work together to calculate the total number of heads and tails tossed by the class. They will use the results on the Penny Flip graph.

3. Guide students in creating a grid in their notebooks to record the outcomes of the experiment. Start by recording the possible outcomes, at the top of the page as follows:

Possible Outcomes for 10 Penny Flips	HT 100	HT 91	HT 82	HT 73	HT 64	HT 55	HT 46	HT 37	HT 28	HT 19	HT 010

4. Below each outcome, record the number of times that outcome was recorded on the graph. For example:

Possible Outcomes for 10 Penny Flips	HT 100	HT 91	HT 82	HT 73	HT 64	HT 55	HT 46	HT 37	HT 28	HT 19	HT 010
# of Times Each Outcome Occurred in Our Experiment	**0**	**0**	**1**	**2**	**10**	**10**	**7**	**4**	**3**	**0**	**0**

5. Below those numbers, separate the number of heads and tails for each outcome. Write the words Heads on one line and below it the word Tails. Going outcome by outcome, determine the number of heads and tails for each outcome. Work together as a class. Here is the start using the same data as above.

Possible Outcomes for 10 Penny Flips	HT 100	HT 91	HT 82	HT 73	HT 64	HT 55	HT 46	HT 37	HT 28	HT 19	HT 010
# of Times Each Outcome Occurred in Our Experiment	**0**	**0**	**1**	**2**	**10**	**10**	**7**	**4**	**3**	**0**	**0**
# of HEADS	0	0	8								
# of TAILS	0	0	2								

6. Have a student explain how to calculate the number of heads for the outcome, 7 H 3 T. Since there were two sticky notes on the graph for this outcome, the total number of heads is calculated by multiplying the number of heads, 7, and the number of tails, 3, by 2.

Possible Outcomes for 10 Penny Flips	HT 100	HT 91	HT 82	HT 73	HT 64	HT 55	HT 46	HT 37	HT 28	HT 19	HT 010
# of Times Each Outcome Occurred in Our Experiment	**0**	**0**	**1**	**2**	**10**	**10**	**7**	**4**	**3**	**0**	**0**
# of HEADS	0	0	8	14							
# of TAILS	0	0	2	6							

7. Have partners determine the number of heads and tails for the 6 H 4 T outcome. Have a student explain how she determined the number of heads and tails. Be sure everyone records the same number for heads and tails for that outcome.

Possible Outcomes for 10 Penny Flips	HT 10 0	HT 9 1	HT 8 2	HT 7 3	HT 6 4	HT 5 5	HT 4 6	HT 3 7	HT 2 8	HT 1 9	HT 0 10
# of Times Each Outcome Occurred in Our Experiment	**0**	**0**	**1**	**2**	**10**	**10**	**7**	**4**	**3**	**0**	**0**
# of HEADS	0	0	8	14	60						
# of TAILS	0	0	2	6	40						

8. Let partners calculate for the other outcomes. For each, have a student tell how he calculated the number. For example, for the outcome, 3 H and 7 T, since that outcome occurred 4 times on the graph, multiply 3 • 4 to determine the number of heads and 7 • 4 to determine the number of tails.

Possible Outcomes for 10 Penny Flips	HT 10 0	HT 9 1	HT 8 2	HT 7 3	HT 6 4	HT 5 5	HT 4 6	HT 3 7	HT 2 8	HT 1 9	HT 0 10
# of Times Each Outcome Occurred in Our Experiment	**0**	**0**	**1**	**2**	**10**	**10**	**7**	**4**	**3**	**0**	**0**
# of HEADS	0	0	8	14	60	50	28	12	6	0	0
# of TAILS	0	0	2	6	40	50	42	28	24	0	0

9. Finally, calculate the total number of heads and tails flipped. Have students use their mental math skills to add the numbers horizontally for heads and tails.

												TOTALS for H&T
# of HEADS	0	0	8	14	60	50	28	12	6	0	0	178
# of TAILS	0	0	2	6	40	50	42	28	24	0	0	192

10. As a check, calculate the total number of heads and tails for each combination of heads and tails.

												TOTALS for H&T
# of HEADS	0	0	8	14	60	50	28	12	6	0	0	178
# of TAILS	0	0	2	6	40	50	42	28	24	0	0	192
	0	0	10	20	100	100	70	40	30	0	0	370

In each case, the grand total adds up to 370 flips of a penny.

11. In this experiment, there were more tails than heads flipped. Have students determine how close these two numbers are to each other. Is there a large difference?

12. Ask students to hypothesize what might happen to the number of heads and tails with additional penny flip data. Allow them time to grapple with this data and its meaning.

Since this is the beginning of the unit, your students may not completely understand the data results. Don't worry! Through the course of the unit they will build an understanding of data and its interpretation.

■ Creating a Class Graph

1. Pose a problem for your students. Ask them to discuss a way to graph the Total Number of Heads and Tails in the Penny Flip Experiment. Have them talk with a partner or in groups.

2. Listen to their suggestions. Put a centimeter grid transparency on the overhead.

3. Have students propose a scale for a bar graph. Attempt to make the graph they suggest on cm grid paper on the overhead. If that scale does not work, ask for another suggestion and try a new scale.

Due to the magnitude of the number of heads and tails, the familiar scale of 1 box to represent 1 head (or tail) in a single column bar graph is not likely to work (unless you have a small class).

4. Once an appropriate scale is determined, complete the graph. Emphasize the importance of labeling its parts. In addition to the scale, the graph needs a *key* to indicate heads and tails as well as include a *title*.

5. Ask students what factual information the graph provides. Listen to their responses and help differentiate between facts and inferences.

6. Ask if the graph gives any indication of the likelihood of getting a head or a tail on a penny flip. Is it equally likely you'll get a head as a tail? Why or why not?

7. Have the students record the total number of heads and tails in their journals. Distribute a sheet of grid paper to each student. For **homework,** using the grid, have students create their own graphs of the total number of heads and tails tossed in the Penny Flip experiment.

Session 4: Graphing Heads and Tails

■ What You Need

- ❑ blank overhead transparencies and pens
- ❑ centimeter (cm) grid transparencies (from Session 3)
- ❑ *(optional)* cm grid paper

■ Getting Ready

1. Gather the overhead grid transparencies used in Session 3. Keep the one with an accurate scale, and clean the rest for use in this session.

2. Decide if you will use the sample graph with an incorrect scale on page 33 to discuss the error. If so, make an overhead transparency of it.

3. Make an overhead transparency of the graph on page 32 comparing motorcycle brands.

4. Collect an assortment of graphs from different sources for your students to read and interpret, appropriate to their skills and abilities. See the Resources section for websites that have graphs.

5. Decide if you will have your student make a second graph of the data after the discussions about scale. Duplicate 1 cm grid per student.

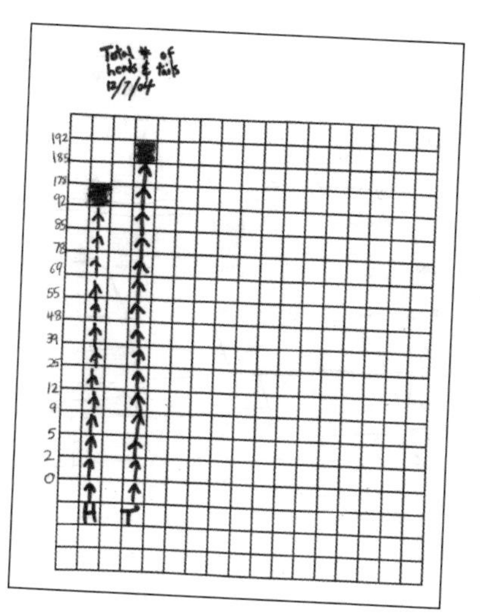

Sample graph with incorrect scale.

GO ■ How Scaling Affects the Graph of the Data

1. Review how the class constructed the bar graph of the data collected from coin flips. Put the class graph from the prior session on the overhead.

2. Have students share the graphs that they made for homework with their table groups or a partner. As students examine one another's work, circulate and note the type of graphs that were made. Pay particular attention to the scales that were created for bar graphs and note any that were inaccurate.

3. When you judge that the groups have completed sharing, ask how many students made a graph with the same scale as was used on the class graph. Look at the data on the class graph and compare it to their work. Is it labeled? Does the bar graph have the number of heads and tails tossed drawn accurately?

4. Ask if anyone used a different scale to make the graph. *For instructional purposes, an incorrect scale is a means for developing deeper understanding of how to create an appropriate scale.*

5. Select a student's graph that you want to discuss. Have that student come to the front of the class. As he explains his scale, record the scale on a grid overhead. Next, have him add the bars to represent the number of heads and tails tossed. For example, here is a the sample graph that a student shared:

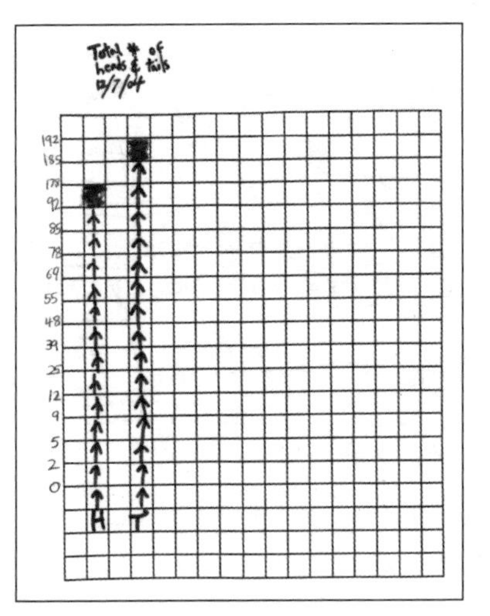

6. Have the students discuss this scale and graph with a partner. Does the scale provide an accurate picture of the data? Why or why not?

7. Emphasize that the unit they select for the scale must be used consistently. Provide a concrete example. Label a graph in increments of two.

8. Ask if this would be a good choice for the scale for the graph of heads and tails. Why or why not?

9. Have another student share her scale for the graph. For example, one student may suggest increments of 20. On a clean grid overhead, use 20 as the unit to make a scale for the graph. It will look like this:

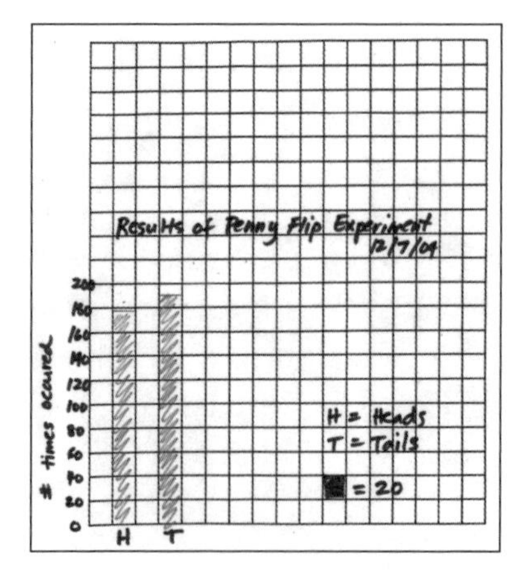

10. Have a student help determine where to mark the number of heads on this graph. Have that student explain how she decided where the number of heads would be located on this scale. Shade the bar to represent heads.

11. Then have another student show where to mark the number of tails. Again, be sure the student explains how he decided where the number of tails would be located on this scale. Shade the bar to represent tails.

12. Ask the students if they think a scale of 100 per unit would represent the data well. Create a graph scaled by 100 per unit. Have students help determine where to mark the number of heads and tails on this graph. What information does this graph provide? Can you see any noticeable difference between heads and tails?

13. Compare the three graphs with scales of 10, 20 and 100. Which one(s) most accurately represents the results of the Penny Flip?

14. Reinforce the idea that the type and amount of data collected informs the scale of a graph.

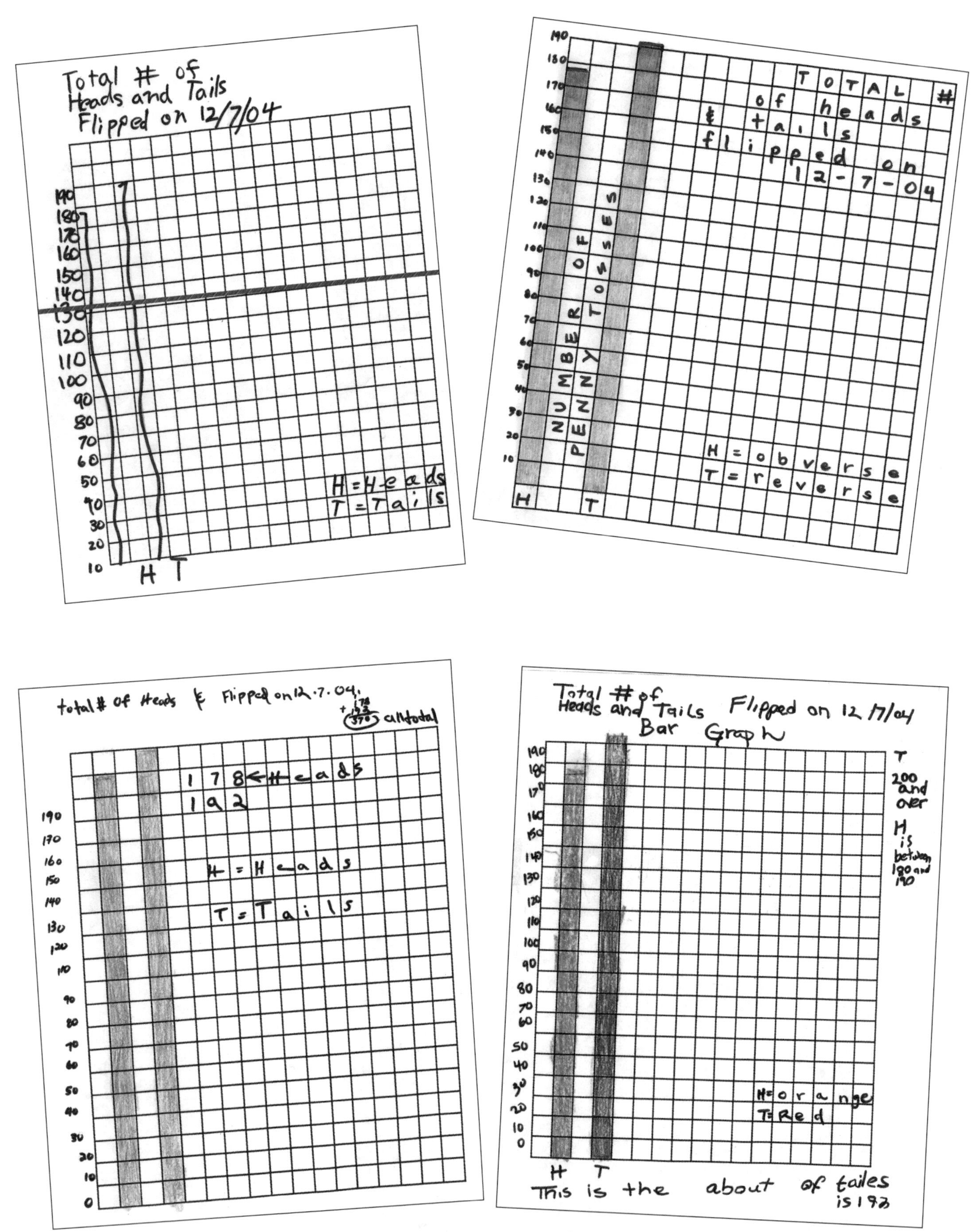

4th grade students' homework samples

■ Truth in Advertising

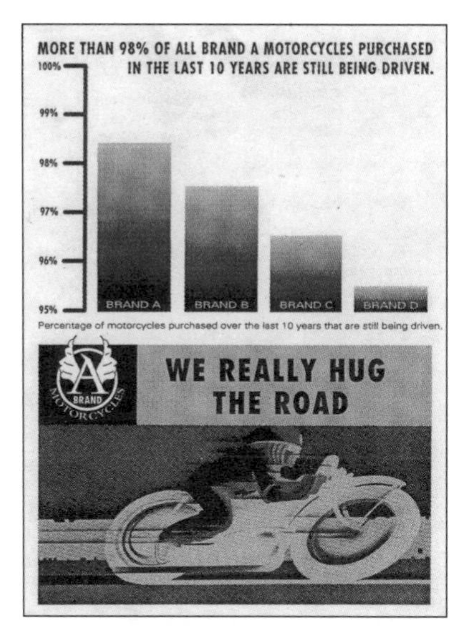

1. Focus the class and put the Motorcycle graph on the overhead.

2. Read the title of the graph together to be sure everyone understands the language on it.

3. Have students discuss the graph with a partner. What information does this graph tell you about the four motorcycles that are being compared?

4. Have a class discussion about their interpretations. It is likely that a student will state that Brand A is better or lasts longer. That is exactly the goal of the graph!

5. Focus the students' attention on the scale. Where does the scale start? Where does it end? How much difference is there among Brands A through D? Why would someone create a graph like this?

6. It is likely you will have a lively discussion once the students realize that the scale of the graph skews the data/information. Emphasize that the intent of the maker(s) of the graph is for you to INFER that Brand A is best.

7. Use a centimeter grid transparency and make another graph of the motorcycle data with your students' help.

8. Have students discuss why it is important to be able to read graphs.

■ Graph Again!

Depending upon your students' skills and abilities, end the session by having them make one additional graph. Below are two options:

Less Experienced Students:
For students who do not have much experience graphing or for those who did not create an accurate graph, guide them into making a labeled bar graph of the Penny Flip experiment on a new sheet of grid paper.

More Experienced Students:
For those students who understand how to make a bar graph and have successfully completed a graph for homework, ask what other type of graph could be used to represent the Penny Flip. For example, have them create a circle graph and compare it to the bar graph. Does one graph provide a clearer picture of the data than the other? Explain why or why not.

■ Homework

Have students look for graphs at home. Ask them to carefully read and analyze the graphs. Note the title of the graph and the SCALE used to display data. Select one graph and provide the following information from the graph:

2 true statements (facts)
1 inference (interpretation of data)
1 question that can be answered by looking at the graph

Session 5: Introducing Theoretical Probability

■ What You Need

- ❑ chart paper, plain (26 x 38 inches)
- ❑ pens

Optional

- ❑ book, *Cloudy with a Chance of Meatballs*
- ❑ string (non-stretchable)

For each student

- ❑ math journals

■ Getting Ready

If you choose not to use this book, begin the class by discussing events in everyday life that always happen and never happen.

1. Obtain a copy of the book, *Cloudy with a Chance of Meatballs* by Judi Barrett (IBSN: 0-689-70749-5). Decide if you will read it to your class.

2. Gather chart paper and pens. If you decide to use a tool to measure the point $^1/_2$ on the 0 to 1 number line, have some string available that is equal to the length of the number line.

GO ■ Always and Never

1. Have students share their homework in small groups. Collect the assignment to assess their understanding of graphs.

   If you decided to do so, read the story *Cloudy with a Chance of Meatballs*. Discuss this tall tale. Is it really possible to have weather come in the form of food?

2. On an erasable board, draw a number line with the benchmarks 0 to 1 as follows:

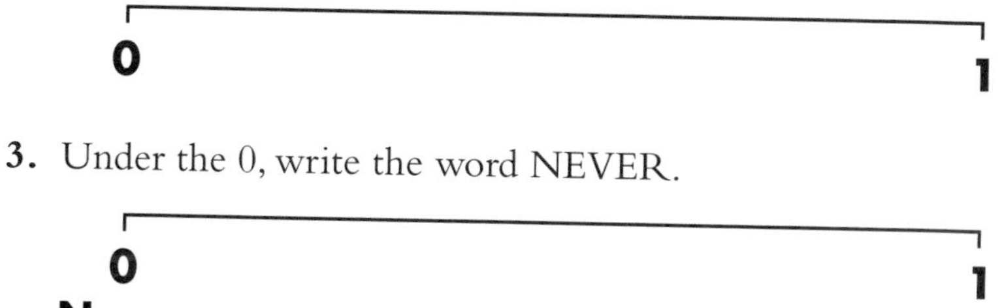

3. Under the 0, write the word NEVER.

0 **1**

Never

Tell students that in this case ZERO represents something that will NEVER occur. It will not happen.

4. Under the 1, write the word ALWAYS.

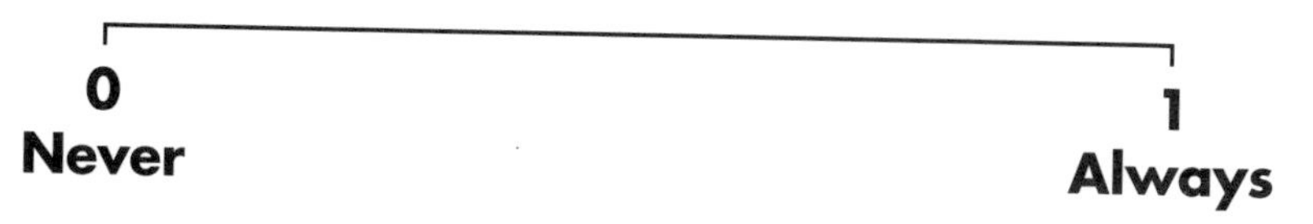

Tell students that ONE represents something that will ALWAYS occur. You can count on it to happen **all** the time.

5. Have students think back to examples of events in the story that would NEVER happen in real life. Have them discuss their ideas with a partner. Record some of their ideas under the zero.

0 **1**

Never **Always**

Humans make useable rafts out of bread.

6. Next, have them think of examples of events in the story that would ALWAYS happen in real life, and discuss them with a partner. Record some of their ideas under the one.

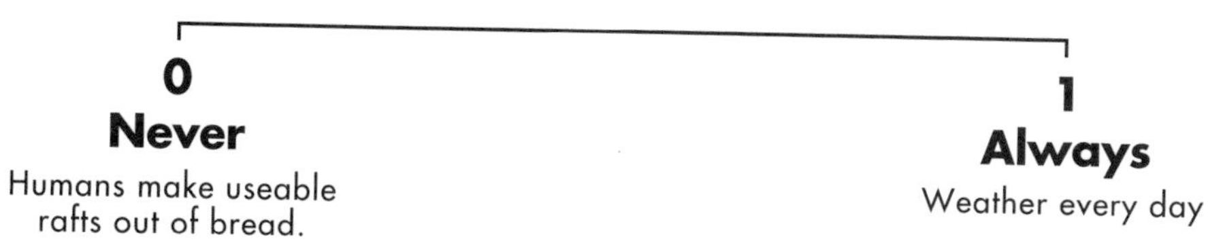

7. Have students think *beyond the story* to other things that NEVER happen and things that ALWAYS happen. Provide time for them to think and record in their journals. After a few minutes, let students talk to a partner and continue recording.

8. Focus the class and ask volunteers to share one item they recorded without telling whether it will always or never happen. After they say their ideas, ask the class where to record them and why.

0 **Never**	1 **Always**
Humans make usable rafts out of bread. A person could go to the Sun and survive. A cat will hatch from a chicken's egg.	Weather every day You need to eat food to live. Earth orbits the sun.

9. Continue and record other ideas on the board. Some ideas students share may occur SOME of the time. In addition, some ideas may be **likely or probable** and others **unlikely or improbable**. Use this vocabulary with your students.

■ Connecting to Probability

1. Have students think back to the Penny Flip. Did each toss ALWAYS come up heads? NEVER come up heads? ALWAYS come up tails? NEVER come up tails?

2. Using their responses to guide your instruction, reexamine the outcomes of a penny toss. There are two possible outcomes—either a head or a tail. Each has an equal chance of being tossed every time the penny is flipped that can be expressed as follows:

$$\frac{\text{1 way to get a head}}{\text{2 total possible outcomes}} \quad \& \quad \frac{\text{1 way to get a tail}}{\text{2 total possible outcomes}}$$

3. You are EQUALLY LIKELY to get a head or a tail on a penny toss. There is a one out of two chance that you get a head or a tail. Mathematically, it can be expressed as the fraction one-half ($^1/_2$).

4. Draw a new number line from 0 to 1 on a sheet of chart paper. Tell students this 0-to-1 scale is a tool that mathematicians use to express the probability of something happening. As in the number line they created about the events in the story, when there is **no chance** of an event ever occurring, it has a **probability of 0**. When an event will **always occur**, it has a **probability of 1**. The number line will look like this:

PROBABILITY OF SOMETHING HAPPENING

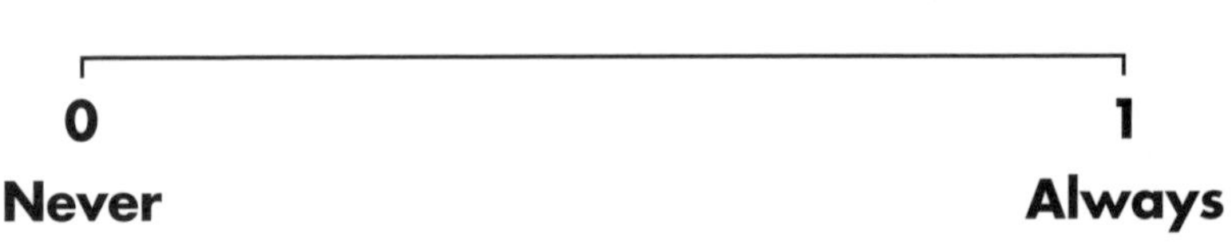

5. Ask where $1/2$ would be situated on the number line from 0 to 1. Have a student come to the board to suggest the location. Check if other students agree. Have other students make suggestions. This point represents an equally likely chance of something happening. Label it on the number line.

To accurately locate the point $1/2$, have a piece of string equal to the length of the line 0–1. Fold the string in half. Use that length to measure the distance $1/2$, and label the point.

PROBABILITY OF SOMETHING HAPPENING

6. Ask if a penny toss is a fair way to decide something. Why or why not? When your class agrees that it is fair, below the $1/2$ write EQUALLY LIKELY. Then add the example of the Penny Flip as follows:

PROBABILITY OF SOMETHING HAPPENING

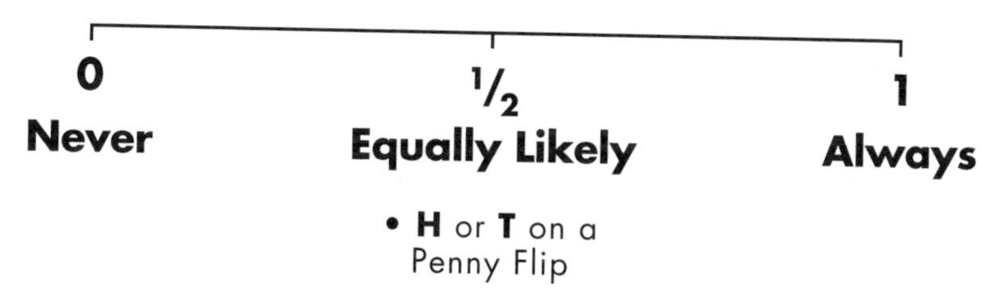

7. Add two more labels to the number line to reinforce the probability vocabulary with a visual representation.

You will be adding to this probability chart throughout the unit. Post it in a location that students can see so you can continue to record on it.

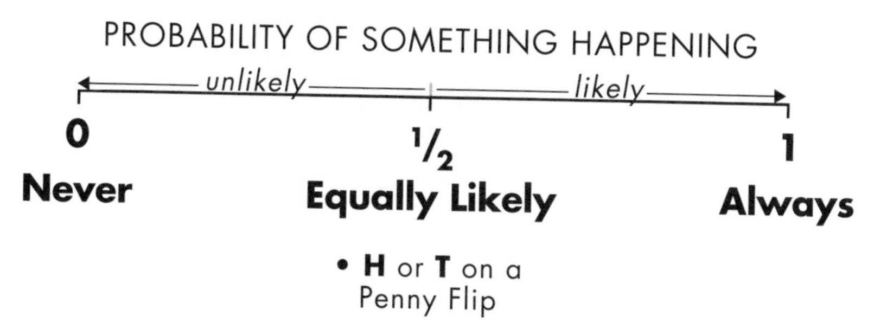

8. Provide a writing prompt to assess your students' understanding of the outcomes of a penny toss and the related probability. Here is a sample:

If you tossed a coin 100★ times, predict how many heads and how many tails you would get.
Write your prediction for 100 tosses: _____ H and _____ T
How did you decide on the number of heads and tails?
Explain your thinking as clearly as you can.

★You can select other numbers such as 250 or 500, depending upon your students' number sense.

9. Collect their journals at the end of the session to assess their understanding and to inform your teaching before you begin the next activity.

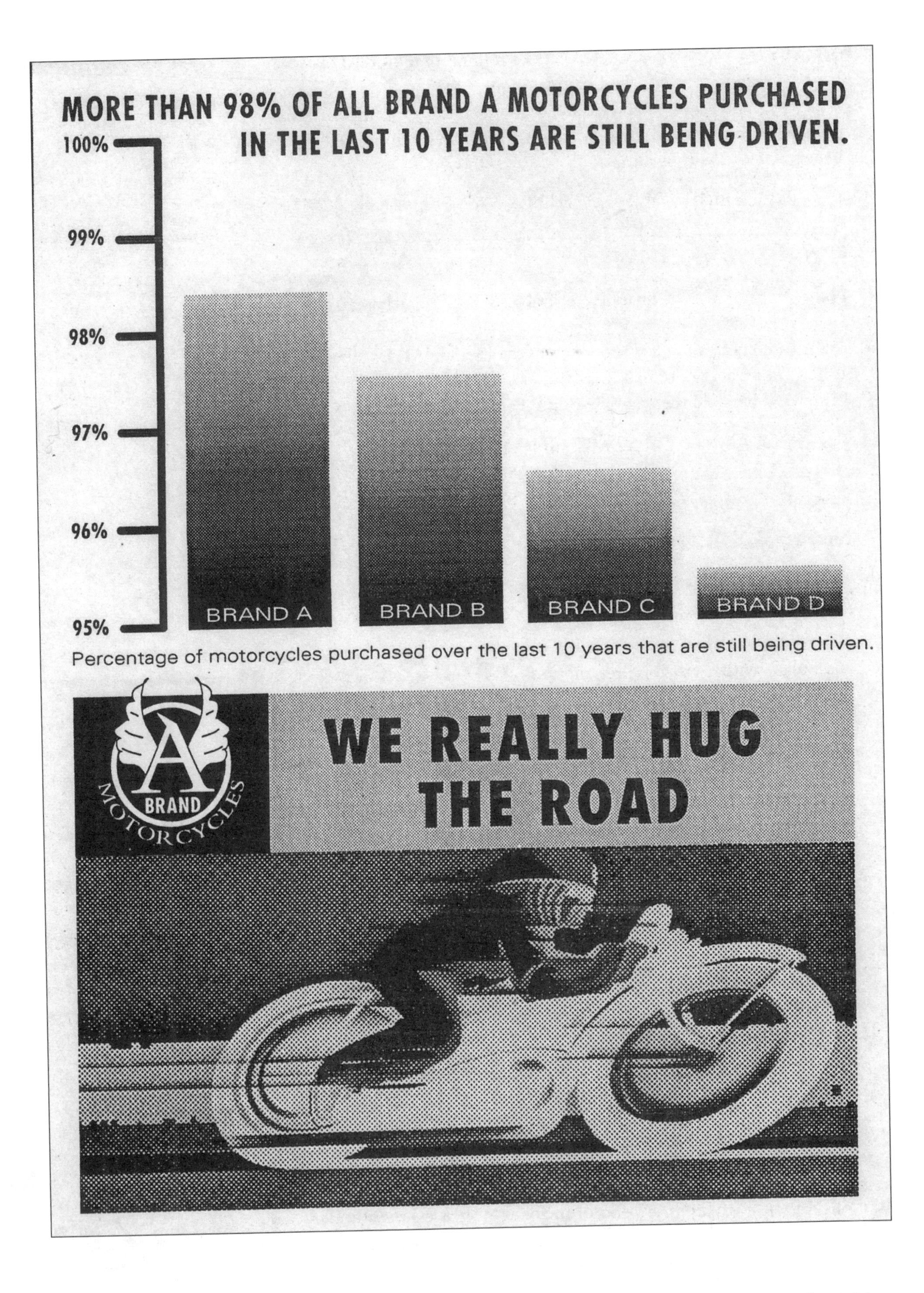
MORE THAN 98% OF ALL BRAND A MOTORCYCLES PURCHASED IN THE LAST 10 YEARS ARE STILL BEING DRIVEN.
100%
99%
98%
97%
96%
95%
BRAND A
BRAND B
BRAND C
BRAND D
Percentage of motorcycles purchased over the last 10 years that are still being driven.
A
BRAND
MOTORCYCLES
WE REALLY HUG THE ROAD

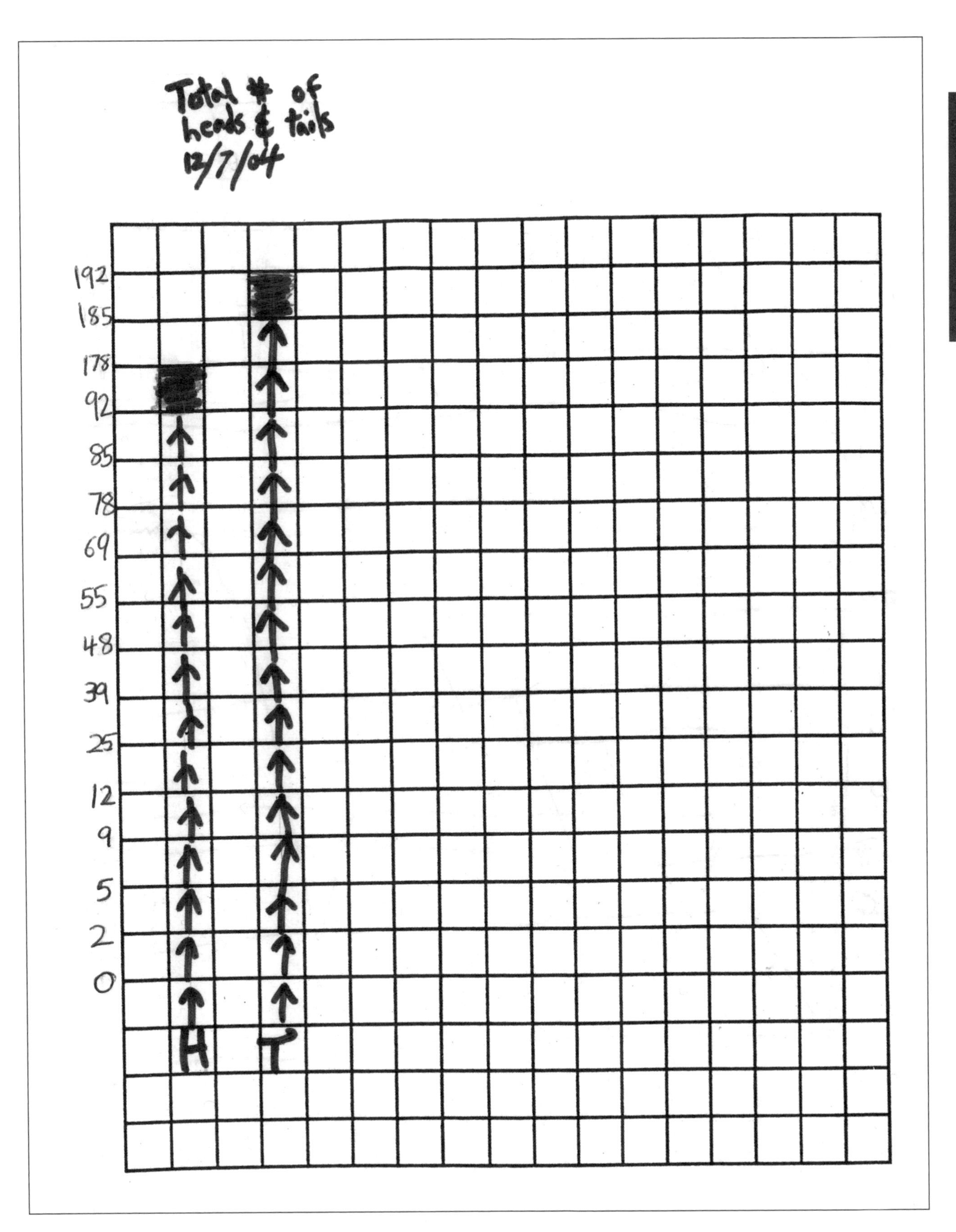
Total # of
heads & tails
12/7/04
192
185
178
92
85
78
69
55
48
39
25
12
9
5
2
0
H
T

STUDENT HANDOUT
ACTIVITY 1

Sample Letter to Families

Dear Families,

Our class is beginning a new mathematics unit called *In All Probability*. As the name suggests, your children will be learning about key concepts of probability. They will also be involved in collecting data, organizing it, representing it, and interpreting it.

Your children already have firsthand experience with probability and statistics in their daily lives. They play games with spinners, coins, cards, and dice that involve probability. They encounter data on food packaging, on television commercials, in news stories, on the sports page, on baseball cards, in political opinion polls, or on a graph about their classmates.

In this unit, children will be involved in four probability experiments. They use pennies, spinners, number cubes (dice), and a set of game sticks to gather, represent, and make sense of the data from the experiments. They will decide whether or not games are fair and the likelihood of events occurring. We will also learn about graphing and examine graphs from newspapers and magazines.

This learning will be "hands-on, minds-on." Students will talk, listen, discuss, read, and write about the mathematics they are learning. They will make predictions and evaluate results. Helping students understand how often and in how many ways their lives and decisions relate to probability and statistics is one of the goals of this unit.

Your children will bring home assignments connected to our work in class. Your family will be asked to look through your pennies and find the two oldest ones. You might be asked to play a game at home and discuss its fairness or help your child find and read graphs. Your children will decorate their game sticks at home for use in the class.

We appreciate your support and "predict" your children will be enthusiastic and interested in probability and data. Chances are that you will want to continue the exploration at home.

Sincerely,

1920
DESIDERO

Overview

This activity uses spinners as the probability tool to generate, organize, and interpret data. Initially, two different spinners are used to play a game. One spinner is divided into equal areas, while the other spinner is divided unfairly. The results of multiple spins are recorded and analyzed using graphs. Finally, a third spinner is presented and students determine its fairness.

In Session 1, students conduct three Track Meet races with each of the two spinners. Spinner B is divided fairly into three equal parts and Spinner A is divided unfairly, in that blue covers $4/8$ (half) the area and yellow and red each cover $2/8$ (one-fourth) of the area. The three colors represent the competing "runners" propelled forward using a spinner. After each race, students record the "winner" on a class graph.

At the end of the third race for each spinner, the final location on the track for each color is recorded on a Track Meet data sheet. The teacher collects these data sheets at the end of the session for use in Session 2.

After the winners of all the races are recorded, students discuss the race results with their classmates, using the class data graphs. They make conjectures about the fairness of the races, focusing on the impact the spinners had on the results. For homework, students revisit and compare the two spinners, and then create two additional spinners—one fair and one unfair.

Session 2 begins with students sharing their fair and unfair spinners from their homework assignment with one another. The homework is collected for teacher review. Optionally, teachers can select an unfair spinner from the assignment for instructional use in Session 3.

Using the data recorded on their Track Meet data sheets, the class generates two bar graphs to compare the distances covered *or the number of spins* for each color. First a "Distance Covered by Runners Using Spinner A" graph is constructed. As students record the number of spins for each color, they see how many times each color was spun. This graph presents a different picture of the race results. Students compare the "winner's graph" with the "distance covered" graph. This provides a more accurate picture of the probability of each color being spun. The same procedure is followed with Spinner B. Then, the students cut a copy of each spinner into its parts to prove that Spinner A is unfair and B is fair.

In theory, the "distance covered" graphs should reflect the percentage of area the three colors occupy on each spinner. In the case of Spinner B, the colors have equal areas. Therefore, the number of spins should be about equal. On the other hand, since blue covers half of the area of Spinner A, the total number of spins for blue should be about half the total spins. Yellow and red should each have about one-fourth of the spins.

After this session, students write a response to a prompt to assess their understanding of each spinner's fractional parts and the impact on its fairness. This informal assessment will help guide the instructional strategies in the next session.

In Session 3, students conduct another experiment based on an unfair spinner. Spinner C is divided into four sections with two of the areas slightly larger than the other two. The spinner is selected to reinforce the concept of a fair spinner; all parts for each color represented must be of equal area! It is not enough for there to be four seemingly equal areas!

Spinner C was designed by a fourth-grade student and was very effective in reinforcing the concept of a fair spinner. Alternatively, you can select a student's spinner from the homework for this activity.

Students use the same procedure for Spinner C as they did with Spinners A and B. First, they conduct three Track Meet Rematch races and record only the winners on a class graph. After the final race ends, the location of each color is recorded on a data sheet. The race is discussed though the lens of the "Winners of Track Meet Using Spinner C."

Finally, students use the data they generated in race three to create a class graph with the distances covered by each color. This provides an opportunity to analyze the spinner through another data lens. Students discuss the two graphs and conjecture if Spinner C is fair or not, based on the data they have generated. As a proof, a model of the spinner is cut into its color pieces and the area of each color is compared. Students concretely see that though the spinner may appear fair, it is not!

To review and assess the concept of a fair spinner, students independently complete a data sheet to evaluate the use of a spinner to make decisions. Their responses are collected and then the sheet is discussed as a class. This provides an assessment of students' individual understanding, as well as the class's understanding, of fairness based on the area of the spinner used to make the decision.

Where's the Math?
This activity provides opportunities for students to conduct probability experiments, gather and organize data on graphs, analyze data, and make conjectures about fairness based on data. Looking at the *number of spins versus the number of wins* demonstrates that the type of data collected provides different insight into determining the fairness of a spinner. The spinners also provide a context for understanding fractional parts of a whole and equivalence of fractions. The concept of "fairness" and related vocabulary are built and reinforced. As appropriate for older students, the theoretical probability for the fractional parts of Spinners A and B is discussed and represented as a fraction between 0 and 1.

Session 1: Ready, Set, Spin!

■ What You Need

- ❑ cardstock, white
- ❑ overhead transparencies
- ❑ chart paper with grids
- ❑ paper clips
- ❑ tape
- ❑ crayons or pens—1 each red, yellow, blue

For each pair of students:

- ❑ Track Meet board
- ❑ 3 markers for track meet—1 each red, yellow, blue
- ❑ 2 Track Meet data sheets
- ❑ 2 spinners
- ❑ crayons—1 each red, yellow, blue

For each student:

- ❑ homework sheet, page 59

■ Getting Ready

1. Use the Track Meet board (master on page 54) to prepare the following:

   a. 1 cardstock copy of the board for each pair of students (non-consumable)
   b. 1 overhead copy of the board for classroom demonstration
   c. 2 copies of the board on white paper to use to record the final race for each spinner (consumable)

2. For each pair of students, make one Spinner A and one Spinner B, using the masters on pages 56 and 57. It is suggested you make one of each spinner first and then decide on a strategy for making the quantity needed.

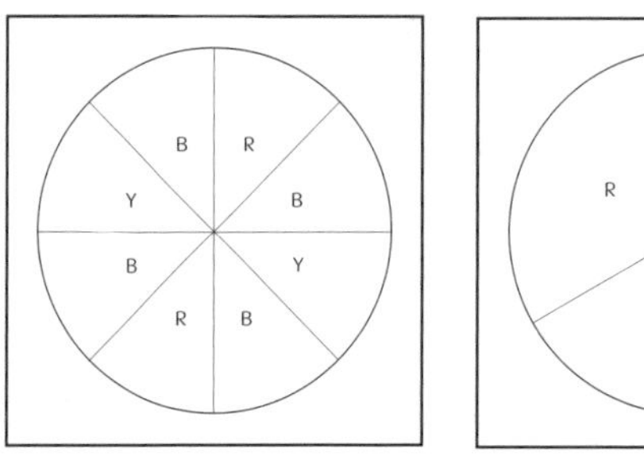

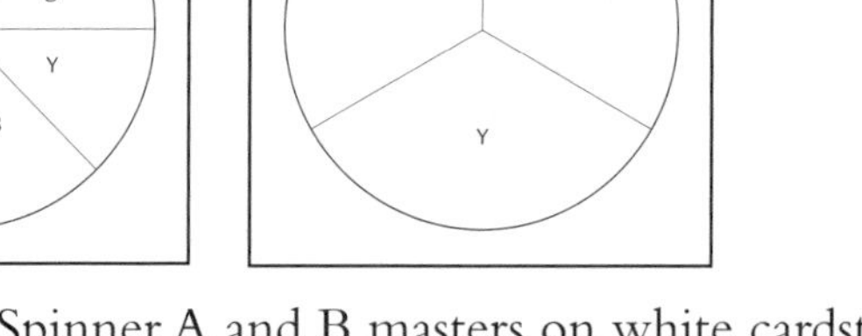

a. Duplicate enough Spinner A and B masters on white cardstock (four/sheet) for your class. Color the circles according to the color labels.
b. Cut the spinners along the dotted lines to create four squares.
c. Unfold a paper clip by bending down the middle section and then straightening it out to form a right angle.
d. Poke the pointed end of the paper clip through the middle of the circle (spinner) from the bottom side.
e. Tape the flat part of the paper clip to the bottom of the spinner.
f. Fold the pointed end of the paper clip down and wrap a small piece of tape around it.
g. Take another paper clip and again bend down the middle section and then make as straight a line as possible with that section.
h. At the other end, create a closed loop. This will go onto the paper clip post.

3. Make an overhead copy of each spinner master (pp. 56 and 57). You will need only one of each. Color each spinner's areas in yellow, red, and blue with permanent markers.

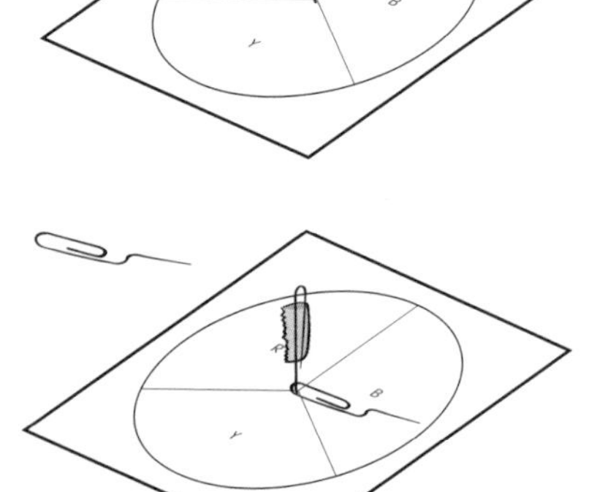

To create the spinner part for this overhead spinner, follow directions 2g and 2h above. Use a sharp pencil to hold the loop in place at the center point of your spinner, and give the paper clip a spin!

4. Gather markers, such as colored plastic disks or tiles, in red, yellow, and blue to represent the racers. Each pair will need three markers, one of each color, for Spinners A and B. You will also need three markers to demonstrate on the overhead.

5. Gather crayons in red, yellow, and blue for students to record the results of race three for Spinners A and B.

6. Using 1-inch grid paper, make two graphs, **one for each spinner,** to record the winners of the Track Meet races. Be sure that the scale for recording the winners is clearly marked and the graphs are labeled, for example:

"Winners of the Track Meet Using Spinner A"

Red																								
Blue																								
Yellow																								

1 box represents one win

For Spinner A, there will be many more wins for blue than for red or yellow. For example, if you have 30 students, there will be 15 pairs and each pair will conduct three races for a class total of 45 wins. Be sure you have at least 30 boxes to record blue winners.

7. Duplicate a copy of the homework sheet on page 59 for each student. Make an overhead transparency copy to explain the assignment to the class.

■ Track Meet

1. Ask the students if they have ever seen or heard of a track meet. After listening to what they know, tell them that they will be simulating a track meet among three "runners."

2. Show the Track Meet board on the overhead. Point out the starting and finishing lines and the runners, identified by red, blue, and yellow markers.

3. Tell the students that the runners move forward according to the outcome on a spinner. Hold up the two spinners. Point out that the three colors are represented on each spinner. Every time the spin lands on a particular color, that runner moves forward one box. The Track Meet ends when one runner has crossed the finish line.

4. Have a student spin Spinner A. Move the appropriate color marker one box forward. Have another student spin. Continue until students are familiar with the game and there is a winner. Be sure to emphasize that the colors or runners are racing—students are not competing with one another.

5. Tell students they will use Spinner A *and* Spinner B for three races each and they will keep track of the winners. They will conduct a **total of six races, three with each spinner.**

6. After the third race using each Spinner, they will "freeze" the runners where they are to record the distance covered by each. Demonstrate how to do this, using the results of the model race on the overhead. Take a copy of the Track Meet board and shade the boxes to show the distance or total number of boxes each runner (color) moved. Use a crayon that matches the color of each runner to shade the boxes.

7. Review the steps for partners to play:

   a. Get a Track Meet game board, three "runners," and Spinner A **or** B.

   b. Conduct three races with that Spinner. Keep track of the color of the winner of each race.

   c. When the third race is over, freeze the runners on the board.

   d. On a copy of the Track Meet board, circle the spinner for that race, either A or B. Record the location of each runner by coloring in the number of boxes they "ran." Set aside.

   e. Trade the spinner for the one you did not use. With the new spinner, follow the same procedure. Have three races, keep track of winners, and record the final race on a copy of the game board.

   f. Distribute student journals as students gather materials.

8. Circulate as students play the Track Meet game and assist them as necessary. Listen to their observations to help inform your debrief of the activity.

9. When students have completed the six races, three with Spinner A and three with Spinner B, focus the class for a discussion. Ask questions to spark their thinking, such as:

   If you were a runner at the Track Meet using Spinner A, what color would you like to be? Why?

   If you were assigned the color Yellow using Spinner B, do you think it is likely that you would win the race? Why or why not?

   Do you think the race was fair for all runners using Spinner A? B? Why or why not?

■ Graphing Track Meet Winners

1. Tell the students that you want to record the winners of the Track Meet races for each spinner. Post the graphs—"Winners of Track Meet Using Spinner A" and "Winners of Track Meet Using Spinner B."

2. First record all the winners for Spinner A. One person from each pair reports the three winners of the races for Spinner A. As they report, make an "X" in the box next to each color to record each win. This will create a bar graph.

"Winners of the Track Meet Using Spinner A"

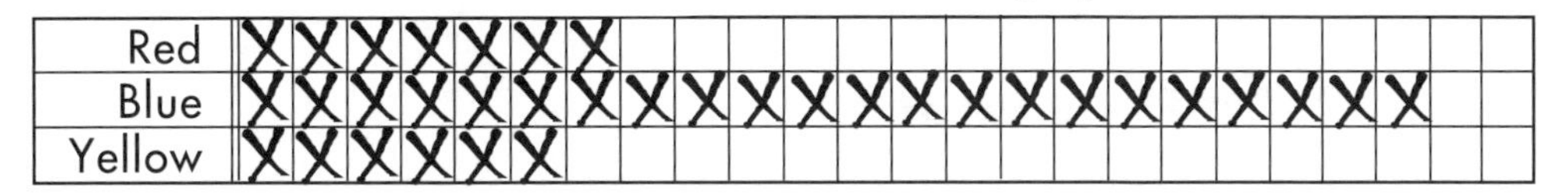

3. As students give results, pause to note the way(s) that the data changes as more information is added. Have partners discuss the graph and share their observations.

4. Next, have the other partner from each pair report the three winners of the races using Spinner B. Record those on the "Winners of Track Meet Using Spinner B" graph.

"Winners of the Track Meet Using Spinner B"

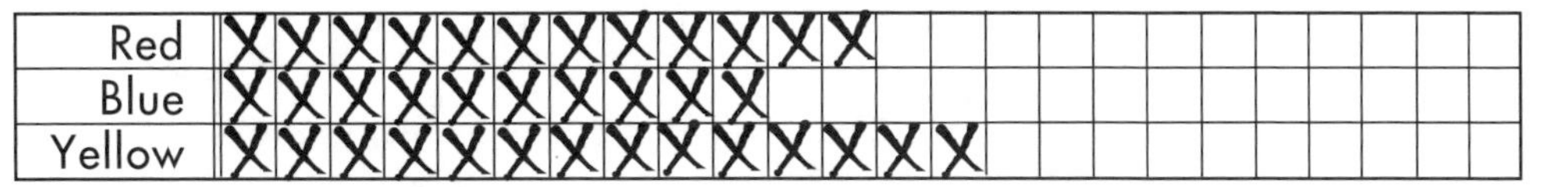

5. Have partners discuss the two graphs. Listen to their observations and ideas about the data.

6. Ask questions about Spinners A and B related to the "fairness" for each runner. Again, first have partners discuss and then have students share their ideas with the class. Come to consensus that, from the results, it appears that Spinner A is unfair and Spinner B is fair.

After looking at the data and re-examining the spinners, students are likely to see that Spinner A is unfair because blue has more area and Spinner B is fair because each color has one-third of the area of the spinner.

7. Place a copy of the homework on the overhead. Point out that it has a copy of Spinners A and B as a reference. There are also two blank spinners. The students' assignment is to create two different spinners—one that they think is FAIR and one that is UNFAIR. They should explain why they think one is fair and the other is unfair.

8. Distribute the assignment and be sure students are clear on what they are meant to do. Collect the Track Meet data that they recorded for the third race for Spinners A and B and have students put other materials away.

Session 2: What's the Real Spin on the Spinners?

■ What You Need

- ❑ grid chart paper with 1-inch squares
- ❑ scissors
- ❑ tape
- ❑ string

For each student:

- ❑ scissors

■ Getting Ready

1. Create **two** graphs using grid chart paper to record where each runner is situated at the end the third race for Spinner A and B. Be sure that the scale for recording the distances is clearly marked and the graphs are labeled, for example:

"Distances Covered by Runners Using Spinner A"

Red																								
Blue																								
Yellow																								

1 box represents one step or one spin

2. Cut the completed Track Meet data sheets from Session 1 into strips by color. There will be three strips per pair of students—one yellow, one blue, and one red **for each spinner.** Have these strips paper-clipped and ready to distribute at the start of the class session.

3. Duplicate the master copies of Spinner A (page 56) and Spinner B (page 57) on regular paper and cut them into four separate spinners. Make enough so that every student has one copy of each and there is one copy of each for you. On your copies, color each section of the spinner. You will post this during the review of the "Winners" graphs made in Session 1.

4. Have crayons available in yellow, red, and blue to color the parts of the spinner. Students will need scissors to cut the paper spinner models.

5. Cut a piece of string the length of the probability number line that was made in Activity 1. It will be used as a tool to locate points on the line.

6. Read the Background for Teachers section on the Law of Large Numbers (page 105) to inform the discussion of Spinner B in the "Parts of a Whole" portion of this session.

7. Decide what prompt you will give your students to assess their understanding of the fairness of the spinners at the end of the session.

■ Winners vs. Distance Covered

1. Start class by providing time for students to share their homework with one another in small groups. Without further class debrief at this time, collect their papers.

2. Focus the students' attention on the graphs of the Track Meet winners using Spinners A and B. Ask what information these graphs provide. Be sure students remember that the graphs show the winners of the races.

3. Post a paper spinner model next to each graph. Have students look at the spinners in relationship to their graphs. Does the spinner help predict the winner? Why or why not?

4. Tell students that you have cut up their data sheets with the results of their third races using Spinners A and B. Distribute the results of Spinner A to one of the partners and the results of Spinner B to the other.

5. Put up the graph titled, "Distances Covered by Runners Using Spinner A." Emphasize that this graph will record the number of moves for *all* of the colors in the race—not just the winners. It will provide evidence of how many times each color was spun for a race.

The graphs for "Distances Covered" will be much longer than for winners. Situate with ample room.

6. Have students come to the graph, one at a time, to tape their results on the graph. Start with the results for red, adding the strips of red one at a time.

"Winners of the Track Meet Using Spinner A"

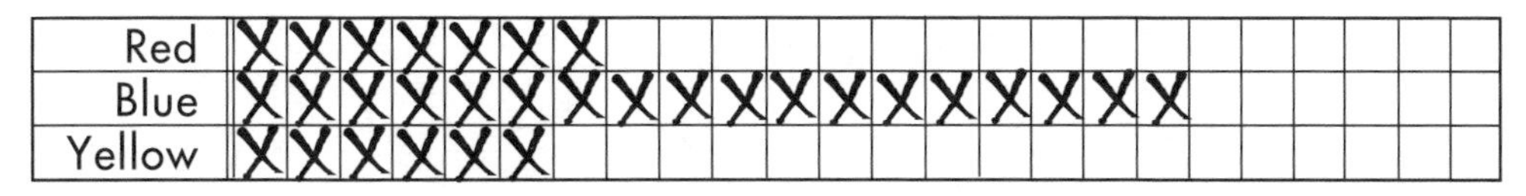

"Distance Covered by Runners Using Spinner A"

Red
Blue
Yellow

In the case of Spinner A, blue clearly had an unfair advantage. However, the number of wins was disproportionate because only winners were recorded. When the ending spot for each of the colors in the race was recorded, the number of spins, versus just the winners, more accurately provides information on the theoretical probability for each color to win.

7. After all the red strips are recorded, compare the results on the "Distance" graph to the results for red on the "Winners" graph. Is there a difference? If so, how have the results changed?

8. Continue using the same procedure and record the results for blue and then for yellow. When all results are recorded for the distances covered, compare the "Distance" graph to the "Winners" graph.

9. Have students talk in pairs or in small groups about the two graphs. Ask for their observations and ideas about the graphs. Pose questions to help them analyze the data, such as:

   What information does each graph provide?
   What does each graph lead one to infer?
   How does each graph relate to the spinner used?

10. Continue by creating a graph for the "Distances Covered by Runners Using Spinner B." Follow the same procedure as for Spinner A.

11. Have students discuss the two graphs for Spinner B. Ask for their observations and ideas about the graphs. Facilitate the discussion, making sure that they explain the reasoning behind their conjectures and ideas.

■ Parts of the Whole

1. Have the students take a closer look at Spinner B. Is it a fair spinner? Why or why not?

2. Distribute copies of the spinner, crayons, and scissors. Have the students color in each section of the circle as labeled by color, and then cut out the pieces.

3. Have them compare the size and shape of each piece. They are all congruent—they have the same shape and size. Therefore, each color has an equally likely chance of being spun using Spinner B.

4. Write the theoretical probability each color represents, as follows:

Probability of spinning a YELLOW Using Spinner B $= \dfrac{\text{1 area of yellow}}{\text{3 equal areas}} = \dfrac{\text{1 chance}}{\text{3 possible}}$

Probability of spinning a RED Using Spinner B $= \dfrac{\text{1 area of red}}{\text{3 equal areas}} = \dfrac{\text{1 chance}}{\text{3 possible}}$

Probability of spinning a BLUE Using Spinner B $= \dfrac{\text{1 area of blue}}{\text{3 equal areas}} = \dfrac{\text{1 chance}}{\text{3 possible}}$

5. Theoretically, Spinner B is fair, even though the results on the graphs may not show that! Tell students about the Law of Large Numbers.

6. Distribute a copy of Spinner A to each student. Have them color in each section of the circle as labeled by color, and then cut out the pieces.

7. Direct students to create a circle by having all the yellow, red, and blue pieces situated together (touching each other).

8. Ask what this new arrangement of the colors tells them about the spinner. Listen to their ideas. This model will show the reason why blue "won" the race more than yellow or red. Blue covers half the area of the spinner!

9. Use this model to discuss probability. Ask students what fractional part of the whole spinner is colored blue. Since blue covers half of the area, what do the students think the chances are for spinning a blue?

10. Continue with the yellow part of Spinner B. Ask what fractional part of the spinner yellow is. Some students may note that yellow is one-half of the remaining half, while others may know it is one-fourth.

11. To make it clear to all students, draw the spinner, arranged in fourths and a half on the board. Divide the half in half, emphasizing that now the whole is divided into four equal parts. Ask what fractional part each color covers and agree that yellow covers $^1/_4$, red covers $^1/_4$, and blue covers $^2/_4$.

12. Have the students place the red and yellow fourths on top of the blue half. This shows the equivalence between $^2/_4$ and $^1/_2$.

13. From these results, ask questions about the likelihood of spin outcomes, such as,

Is it likely, unlikely, or equally likely to spin a blue or a red?
Is it likely, unlikely, or equally likely to spin a yellow or a blue?
Is it likely, unlikely, or equally likely to spin a yellow or a red?

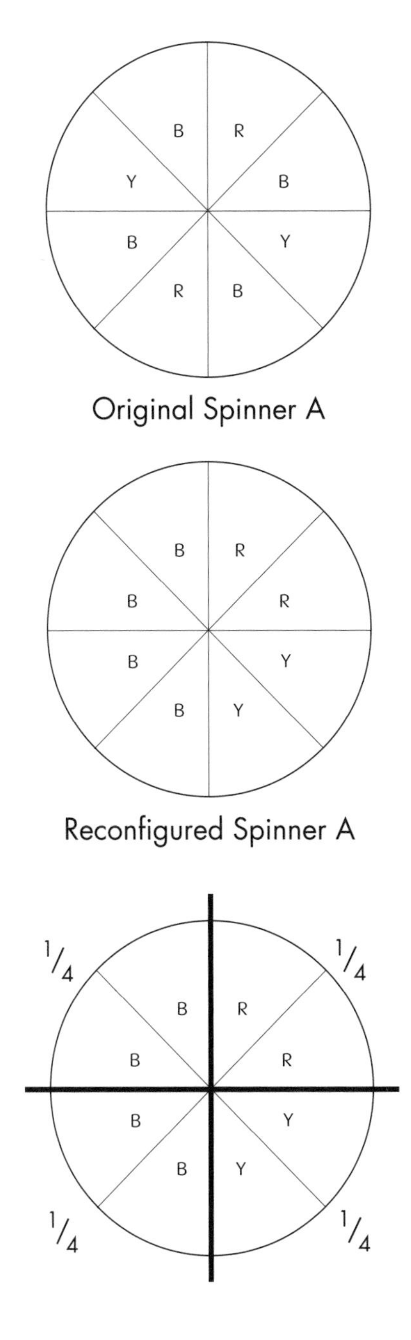

14. Write these fractions in terms of the theoretical probability each represents, as follows:

Probability of spinning a YELLOW Using Spinner A = $\frac{\text{1 area of yellow}}{\text{4 equal areas}}$ = $\frac{\text{1 chance}}{\text{4 possible}}$

Probability of spinning a RED Using Spinner A = $\frac{\text{1 area of red}}{\text{4 equal areas}}$ = $\frac{\text{1 chance}}{\text{4 possible}}$

Probability of spinning a BLUE Using Spinner A = $\frac{\text{2 areas of blue}}{\text{4 equal areas}}$ = $\frac{\text{2 chances}}{\text{4 possible}}$

In terms of the outcomes, blue has a 2-out-of-4 chance of being spun, and because $^1/_2$ is equivalent to $^2/_4$, it can also be expressed as a 1-out-of-2 chance of being spun.

■ Theoretical Probability (for Students in Grades 4 and 5)

1. Link the discussion on the fractional probability representations to the "Probability of Something Happening" chart in Activity 1, Session 5. Tell students you want to put the chances of spinning each color on Spinner A on the chart.

2. Ask where to situate the probability of spinning a blue with Spinner A. Listen to ideas and agree it would be at $^1/_2$, since it is equally likely to get a blue as a red or a yellow combined. Label the point one-half with the equivalent fraction, $^2/_4$, and write "Blue on Spinner A" below.

3. Ask where to situate spinning a red. Listen to responses and agree it would be at the point $^1/_4$. Ask a student to estimate where $^1/_4$ is situated on the number line.

4. Use string to locate the point more precisely. Show students that the length of the string equals the distance 0 to 1. Fold it in half and measure the distance to $^1/_2$ with the string. Fold the string in half again. Ask how many equal parts the string is divided into. [4] Each part is $^1/_4$ and can be used to locate the point $^1/_4$. Use the string to measure and mark the point $^1/_4$. Label it "Red on Spinner A."

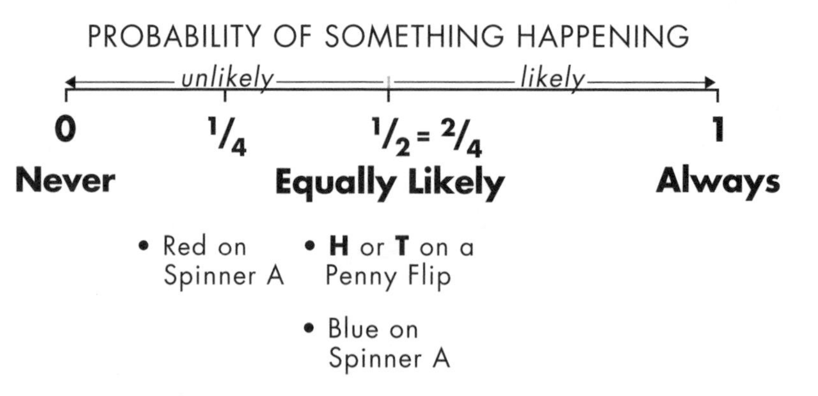

5. Finally ask students where to situate the probability of spinning a yellow. Have a student go to the chart, locate the point, and explain why it was chosen. Record it on the chart.

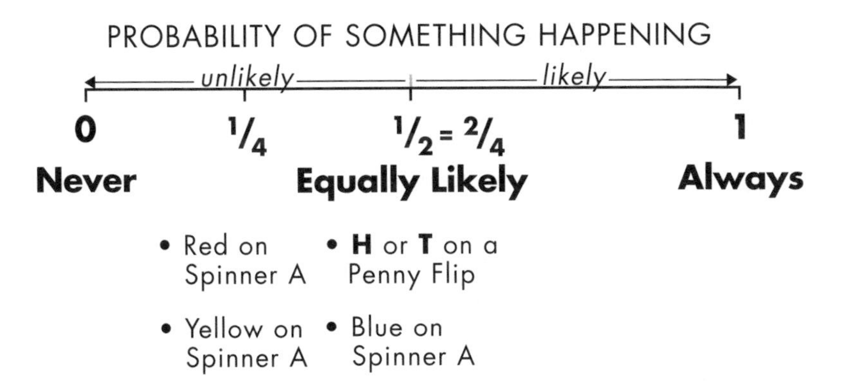

Note that even though it is unlikely to get a yellow or a red, those colors are sometimes spun, but not as frequently as blue.

■ Writing Prompt (for Grades 3–5)

End the session with a writing prompt to be completed in class or for homework to help you assess student understanding at this point in the unit. Here are some sample prompts:

Related to Spinner A:

A friend comes to you for advice. In her family, the three children take turns doing the dishes. For this week, the children decided to use Spinner A to determine which child will do the dishes for the week. Each child will be represented by one of the colors—red, blue, or yellow—on the spinner. Your friend does NOT want to do the dishes.

What color would you recommend your friend select? Tell the color and say why that color is a good choice.
Explain your reasoning to your friend in as many ways as possible, using models, diagrams, and words.

Related to Spinner B:

If you were playing the Track Meet game with Spinner B and you were given the color RED, do you have a FAIR chance of winning the Track Meet race? Why or why not?

Explain your answer in as many ways as possible using models, diagrams, and words.

Session 3: Spinner Investigation: Fair or Not Fair?

■ What You Need

- ❑ cardstock, white
- ❑ overhead transparencies
- ❑ chart paper with 1-inch grid
- ❑ paper clips
- ❑ tape
- ❑ crayons and pens
- ❑ 1 copy of Spinner C

For each pair of students:

- ❑ Track Meet Rematch board
- ❑ markers—1 each yellow, blue, red, green
- ❑ crayons—1 each yellow, blue, red, green
- ❑ Track Meet Rematch data sheet
- ❑ 2 copies of Spinner C (one for each student)
- ❑ 2 copies of *Fair or Not Fair* Assessment (one for each student)

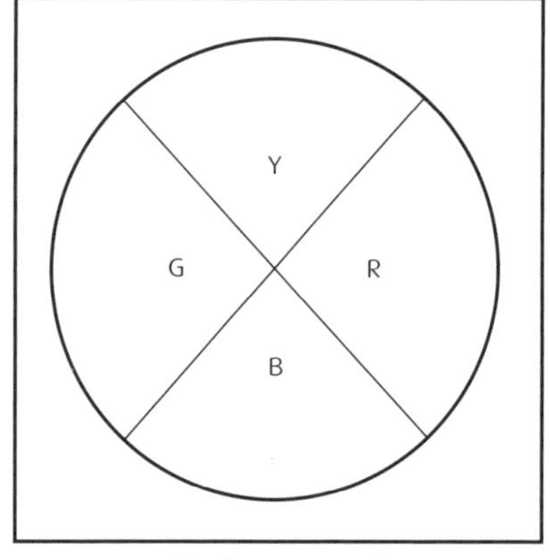

Spinner C

Alternatively, you may choose to select one of your students' spinner designs as the spinner to test for fairness in this Session. If so, you will need to create a master copy of the spinner as well as a Track Meet Rematch board to prepare the materials needed for the session.

■ Getting Ready

1. Review your students' homework from Session 1 to determine how well they understood the concept of fair and unfair spinners. This will help identify which students may need extra support as you present this session and inform your instruction for the entire class.

2. Use the Track Meet Rematch master on page 55 to prepare

    a. 1 cardstock copy per pair of students (non-consumable)
    b. 1 paper copy for each pair of students (consumable)
    c. 1 overhead copy of the board for demonstration

3. Using the Spinner C master on page 58,

    a. Duplicate enough spinners on cardstock so that you have one per pair of students (non-consumable)
    - Color the spinners according to the labels.
    - Follow the procedure for making spinners as described in step 2 of the Getting Ready section of Session 1 (page 40).

    b. Duplicate enough copies on regular 8.5 x 11 paper so that each student and you will have one copy of the spinner to cut for comparing its parts. (consumable)

4. Gather the game markers used for the Track Meet in Session 1 and an additional green marker for each pair of students.

5. Gather the crayons used in Session 2 to color spinners and an additional green crayon for each pair of students. Have scissors available for the students.

6. Create two graphs on 1-inch grid chart paper to record the winners of the races and the distances covered (as you did for Spinners A and B in Sessions 1 and 2).

"Winners of the Track Meet Rematch"

Yellow																								
Blue																								
Red																								
Green																								

1 box represents one win

"Distances Covered by Runners in the Rematch"

Yellow																								
Blue																								
Red																								
Green																								

1 box represents one step or one spin

7. Use the *Fair or Not Fair?* assessment on page 60 to prepare:

   a. 1 copy per student to complete independently in class
   b. 1 overhead transparency to debrief and discuss the spinner with the class
   c. 1 paper copy of spinner to cut into parts, as needed during discussion

GO ■ Track Meet Rematch

1. Tell students that you have a new spinner and you need their help to determine if it is a FAIR or an UNFAIR spinner.

If you use one of your students' spinners, let that student know ahead of time and ask her not to tell if it is fair or unfair!

2. Show the spinner in question on the overhead. Have students discuss the spinner's fairness with a partner or in small groups.

3. Show the corresponding board, Track Meet Rematch, to use with the new spinner. Point out that now there are four colors racing.

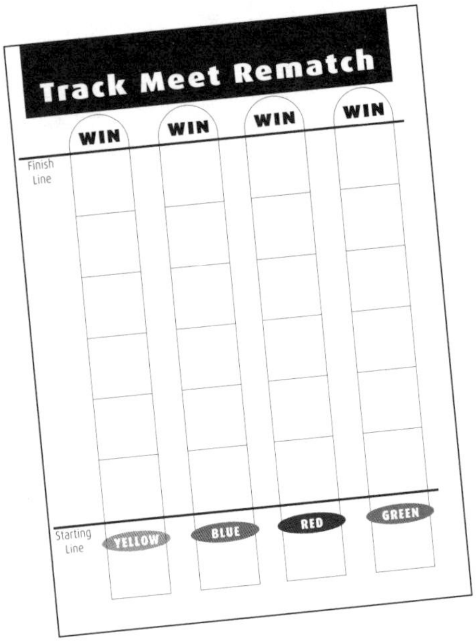

4. Tell students they will use the same procedure for this spinner as they did for Spinners A and B. Review the procedure:

   a. Conduct the race three times.

   b. Record the winner of each race on the class chart "Winners of the Track Meet Rematch".

   c. At the end of the third race, freeze the runners.

   d. Use a data sheet to record the placement of each runner at the end of race three.

   Have a student repeat the directions for further clarification and to support all students' understanding.

5. Distribute a spinner, a Track Meet Rematch board, a Track Meet Rematch data sheet, and markers to each pair of students.

6. Circulate as students play and record the winners of each race. Be sure they record the ending spot for each racer in the third race accurately on their Track Meet Rematch data sheets.

7. When all the races are complete, focus the class on the "Winners" graph. From the data, what can they conjecture about the spinner? Is it fair? Unfair? Why or why not?

8. Let students discuss among themselves first, and then facilitate the discourse to hear their thinking about this spinner.

9. Post the "Distances Covered by Runners in the Rematch" graph. Instead of cutting up strips for each color, have students report the number of spins for each color and then you record the data on the graph. Record in one box for each spin.

10. Start with a yellow and have students report the number of spaces yellow moved. Record in the corresponding number of boxes on the class graph. Continue color by color until the graph is complete.

11. After the graph of the distances covered is complete, ask if this data provides any new insight into whether or not the spinner is fair for this race. Why is it fair or why isn't it fair?

12. Compare the two graphs about data generated from the spinner. Ask for student ideas and observations. Pose questions to spark ideas as necessary, such as:

How is the data on the graphs related?
Which graph is more helpful in determining whether it is a fair spinner? Why?
Does the "distances covered" graph provide information on which color(s) is (are) most likely to be spun? Color(s) least likely to be spun?

13. Provide a tool to concretely determine if the spinner is fair. Distribute a copy of the spinner to each student. Have them cut the spinner into its color pieces and compare the size of the pieces. Ask if all the color pieces have equal area. Do any have equal area? If so, which ones? (Red and green are equal; yellow and blue are equal.)

14. The model concretely proves that the spinner is unfair! There are two colors (red and green) that have a greater area than yellow and blue. Connect the spinners' fractional parts back to the results on the graphs. Does the data correspond to the areas of each color on the spinner? For example, since red has a larger area than blue, did red have more wins and cover more distance?

■ Assessing Understanding

1. To assess the conceptual understanding of a fair spinner, distribute the *Fair or Not Fair* data sheet for students to complete independently.

2. Provide about five minutes for students to work and then collect papers. Their individual responses will inform your instruction.

3. Put the overhead copy of the *Fair or Not Fair* data sheet on the overhead. Have students talk to a partner or in small groups about the fairness of the spinner for making decisions.

4. As a class, discuss the use of the spinner to make each decision. Have students debate and make convincing arguments about the fairness of the spinner for Decision 1 and 2. For Decision 1, since the spinner is devided into seven equal areas and each of seven children have one area, the use of spinner is fair. However, since there are four girls and only three boys, the girls have more area on the spinner, and it is not fair to use it in this case.

Have a copy of the spinner available to cut for making proofs.

Track Meet

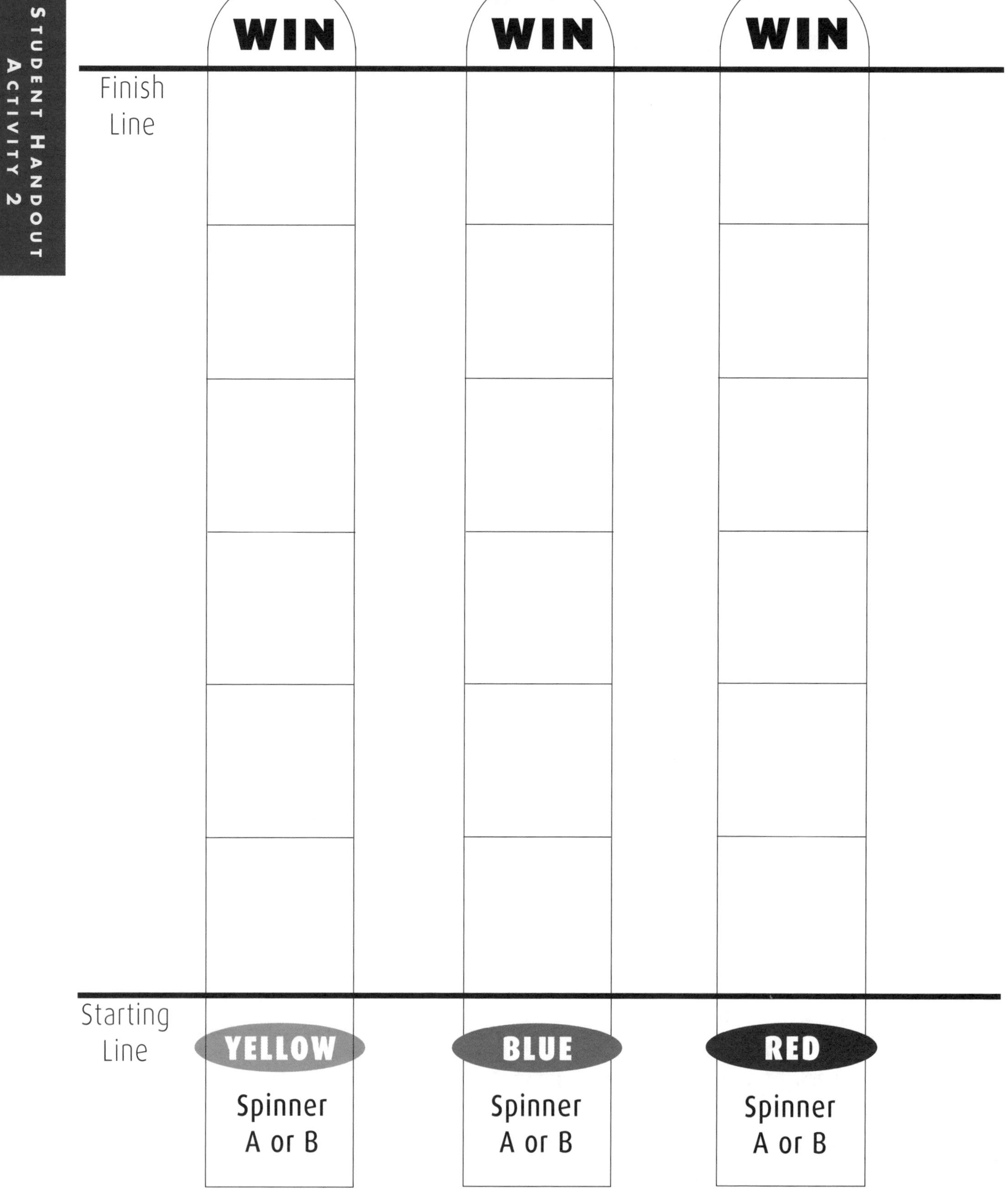
WIN
WIN
WIN
Finish Line
Starting Line
YELLOW
Spinner A or B
BLUE
Spinner A or B
RED
Spinner A or B

WIN	WIN	WIN	WIN

Finish Line

Starting Line

YELLOW	BLUE	RED	GREEN

Spinner A

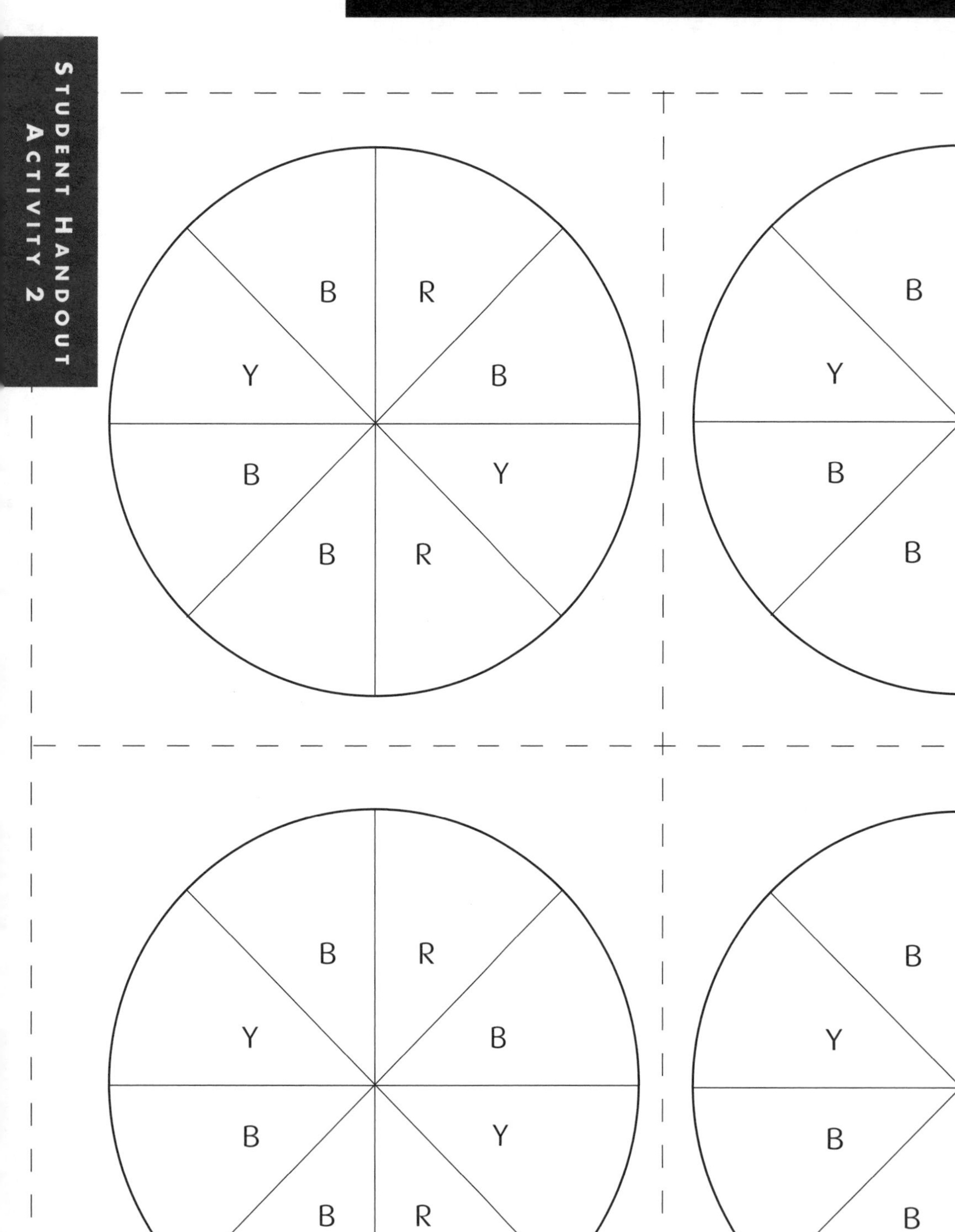

Spinner B

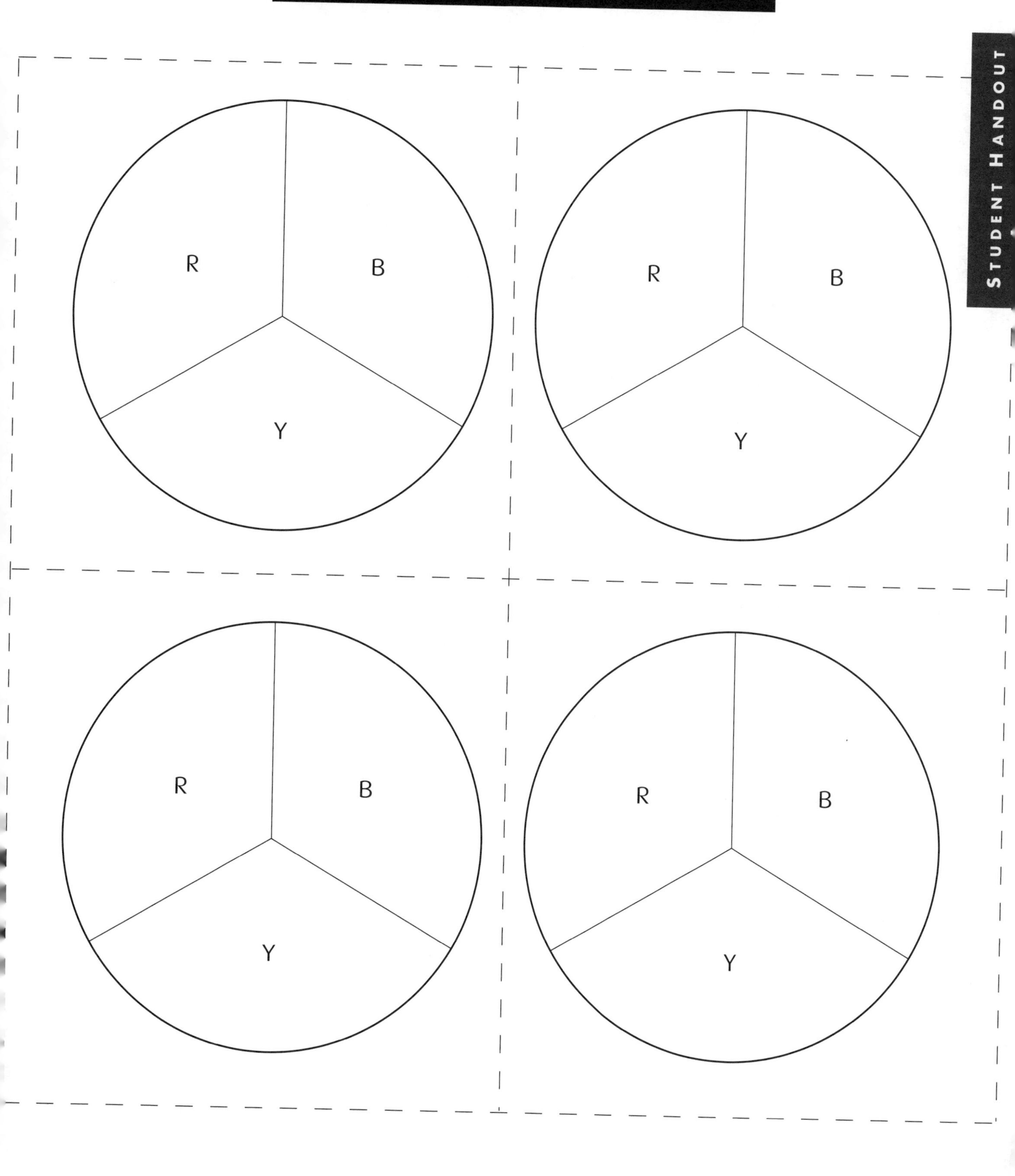

Spinner C

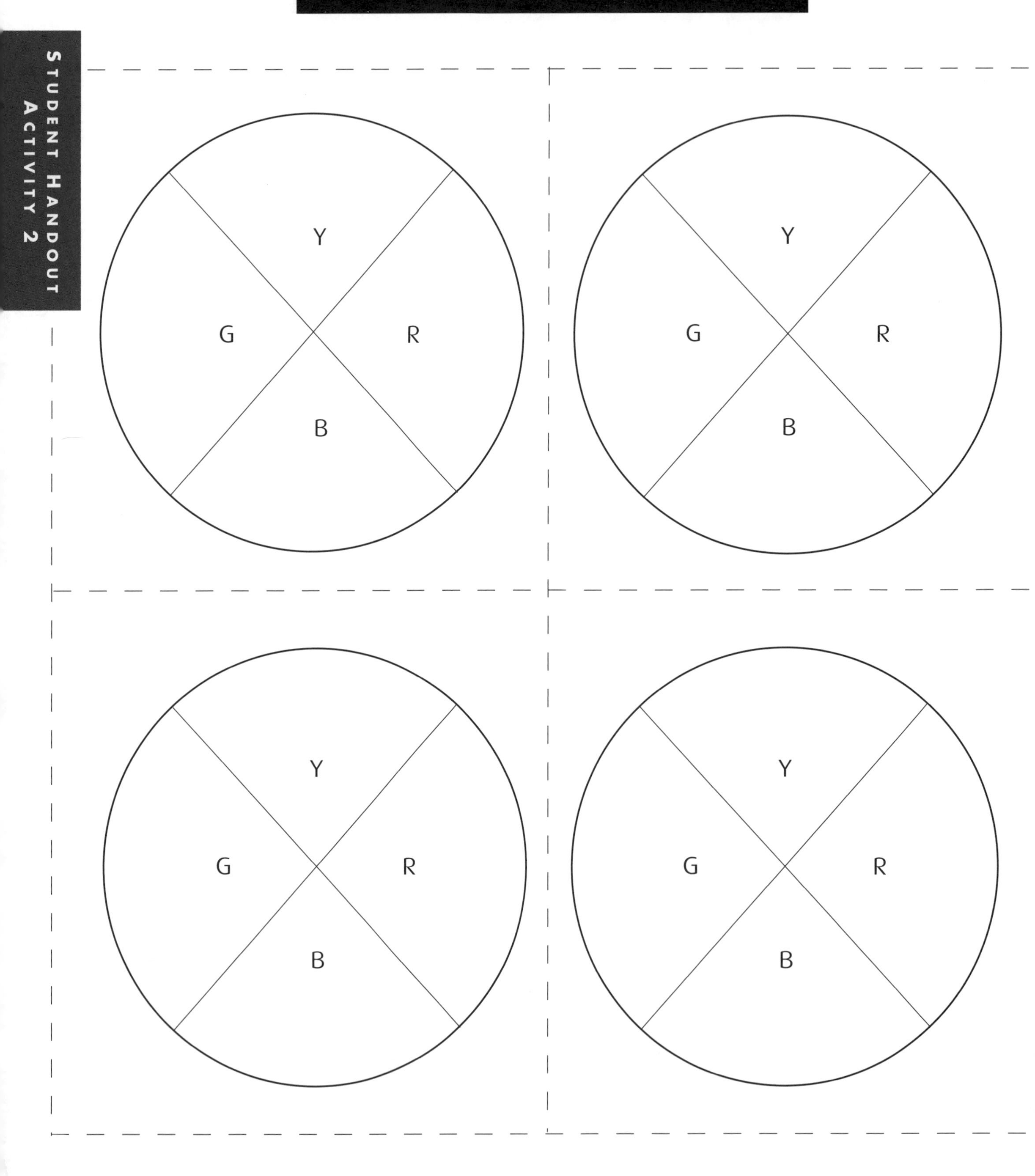

Today we played Track Meet using these 2 spinners:

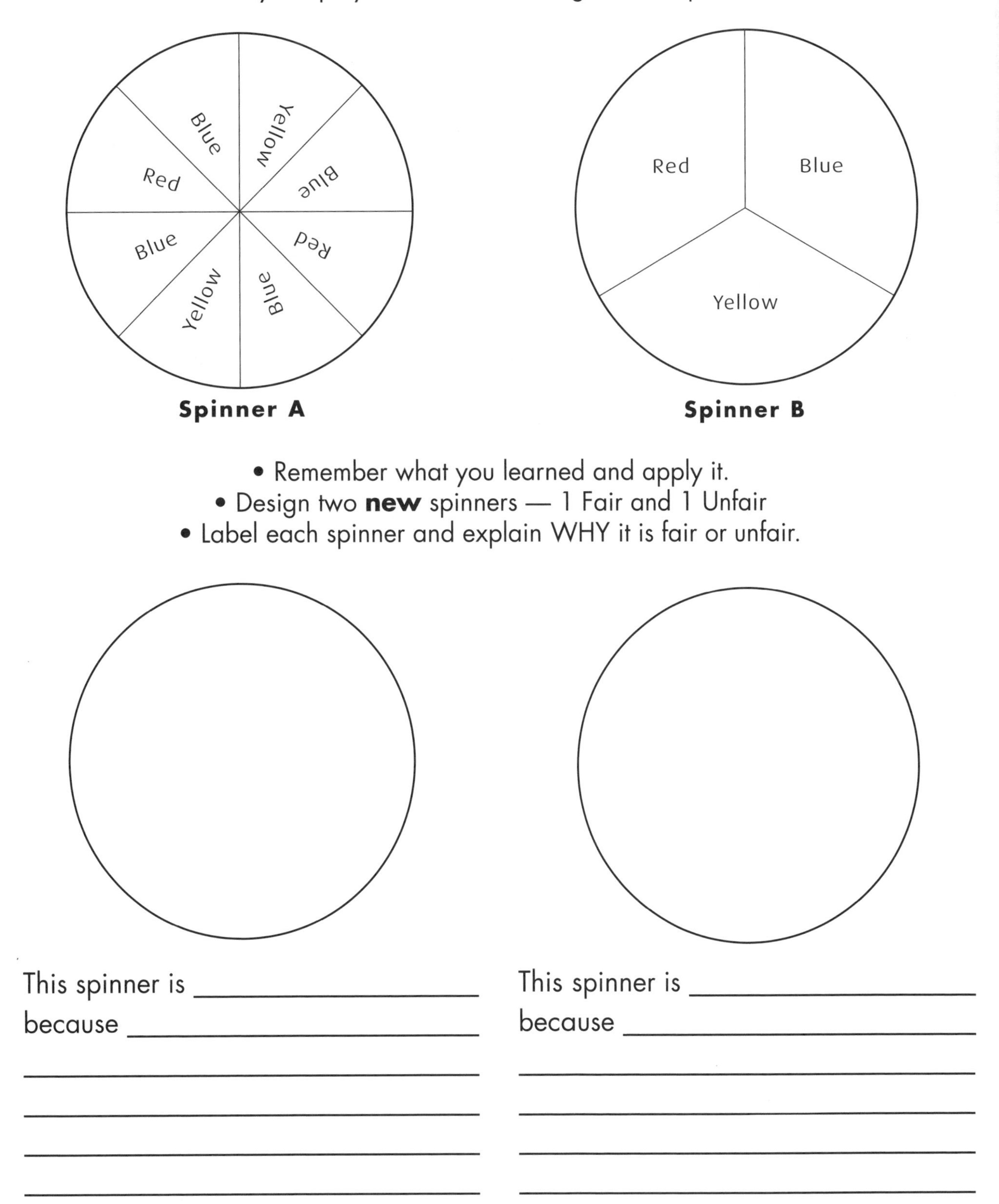

- Remember what you learned and apply it.
- Design two **new** spinners — 1 Fair and 1 Unfair
- Label each spinner and explain WHY it is fair or unfair.

This spinner is ______________________
because ______________________
____
____
____
____

This spinner is ______________________
because ______________________
____
____
____
____

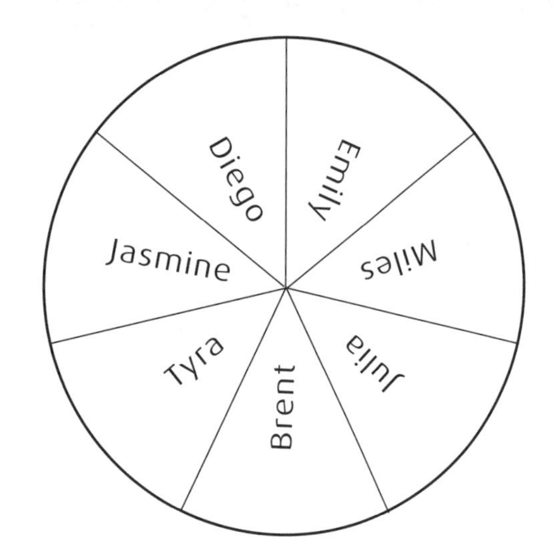

This spinner is divided into equal parts.

A group of friends who play together use it to help them make decisions.

Each person wrote his or her name in one part of the spinner.

They want you to help them understand if the spinner is fair for making the following two decisions.

Decision 1: Everyone wants to go first in a board game. If they use the spinner to decide who should go first, does each person have a FAIR chance to go first?

YES NO

Explain why:

Decision 2: The friends decide to play a game where the girls will be on one team and the boys will be on the other team. They need to decide who will go first. Is the spinner a FAIR way to decide?

YES NO

Explain why:

Overview

In this activity, students use standard dice to conduct probability experiments and gather data. They begin by sharing their knowledge and experience with dice, and use a single die to determine its "fairness." Then they conduct a horse race, the Double Dice Derby, to investigate the outcomes when rolling two dice.

In the first session, students begin with the Horse Race game or with the activity Roll ALL Six. In each, they use a single die to generate numbers. In the Horse Race game, horses, named by the numbers one through six, move forward one space according to the number of each roll of the die. The race ends when one horse makes it into the "winner's circle." After students complete two races, the winners of each race are recorded on a class chart and discussed. Then students race the horses two more times, again record the winners, and observe how the data changes with more results included. As time permits, the horses race again and additional results are recorded.

Most third- and some fourth-grade teachers found that the Horse Race worked best to investigate the outcomes of one die. Depending upon your students' skills and abilities, select the appropriate activity for your class.

Alternatively, students conduct a Roll ALL Six experiment in which they roll a die until each of the numbers on the die is rolled at least one time. They create their own data sheets and work with a partner to record the number of rolls. The class results are gathered on a chart and students compute the total number of rolls for each number on the die. Then the data is analyzed and students make conjectures about how fair a roll of a single die is.

In each of the opening activities in Session 1, connections are made to the prior activity, in which a spinner had three equal areas, and to the Penny Flip activity. As appropriate, the probability is expressed as a fraction and situated on the scale of zero (never happen) to one (always happen).

Session 2 begins with a lively horse race, the Double Dice Derby, with 12 horses. The "competing" horses race forward, step-by-step, according to the sum of each roll of two dice. The race ends when one horse makes it to the "winner's circle." As the students play, there is a lot of excitement in the room. They record their winners on a class data chart, and each pair plays at least four games.

As they play, students are likely to observe how the dice sums impact the ability of a particular horse to win. For example, Horse Number 1 never moves! And Horses 6, 7, and 8 seem to win a disproportionate amount of times. After the games, a discussion about the data provides an opportunity to begin to make sense of the race results. For homework, the students take home a Horse Race board and play the game at least three times and record the results.

Session 3 opens by recording the results of the horse races students completed for homework. After their data is added, the students discuss how the additional race results impact the data recorded the previous day. The data gathered is converted to a line graph to provide another picture of the results.

After this discussion, students complete a "Keeping Track" chart to determine all the possible sums for two dice in an organized manner. The chart provides a concrete picture of why certain horses win more frequently than others. Rich discussions of the patterns on the chart lead to deeper understanding of the outcomes and the reason why the race is not fair!

A line graph can also be connected to the "Keeping Track" chart. If appropriate for your students' abilities, the probability for each sum of two dice is recorded as a fraction and situated on a scale of zero (never happen) to one (always happen).

At the end of Session 3, students respond to a writing prompt to assess their understanding of the probability of rolling each of the possible sums for two standard dice.

Where's the Math?

The Horse Races and Roll ALL Six activities provide another context for gathering data through probability experiments. Data collection and organization play a critical role in understanding the outcomes for a single die roll or for the sum of two dice. A line graph is introduced as a tool to evaluate the data through another lens. Students make conjectures about the "fairness" of the outcomes in all of the experiments and use data to justify their thinking. Computational practice is built into the activities through the data collection and analysis. The outcomes of rolling a die and two dice are expressed as fractions. The theoretical probability of the outcomes can be expressed on the number line from zero to one, depending on students' understanding.

Session 1: Roll of a Die

■ What You Need

Materials:

- ❑ cardstock
- ❑ overhead transparencies
- ❑ overhead pens
- ❑ 1 die
- ❑ 6 markers (beans, cm cubes, plastic circles)
- ❑ chart paper with 1-inch grids
- ❑ pens, 2 different dark colors

For each pair of students:

- ❑ 1 die
- ❑ 1 Horse Race board
- ❑ 6 markers (beans, cm cubes, plastic circles)

For each student:

- ❑ math journal

Optional:

- ❑ 1 giant-sized pair of standard dice

■ Getting Ready

1. For Sessions 1, 2, and 3 you will need at least two dice per pair of students. We strongly recommend that the pair of dice be two different colors to help students see that the roll Green 3 and White 4 is distinct from Green 4 and White 3. You many also want one pair of large dice for group demonstrations (See page 135 for sources). Gather enough dice at the onset of this activity so that the dice will be available throughout.

2. Read through both probability experiments, Horse Race and Roll ALL Six, and decide which of the two experiments you will do with your students. Follow the preparation for the one you choose.

3. If you decide to use the Horse Race prepare the following:

   a. Using the Horse Race board (master on page 85),
      - Make one cardstock copy of the Horse Race for each pair of students.
      - Make an overhead transparency of the Horse Race to introduce it to the class.

   b. Gather one die and six game markers per pair of students and for yourself.

c. Make a graph on chart paper with 1-inch grids to record the class results of the horse races as follows:

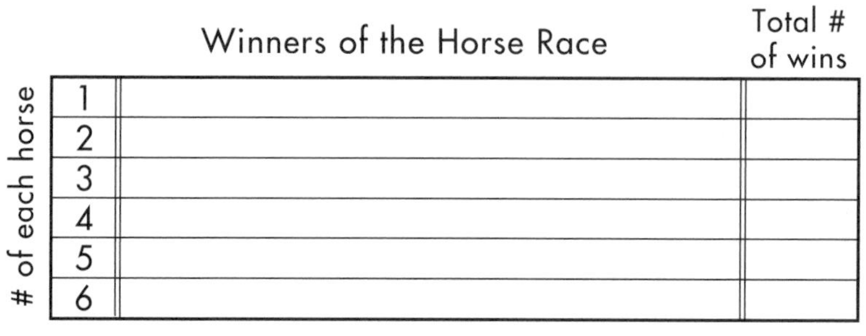

1 tally represents 1 win

4. If you decide to use the Roll ALL Six experiment, make a chart to record the Roll ALL Six class results. The data you collect will be the total number of times each of the numbers on a single die (one through six) were rolled by each pair of students. Be sure to label the chart. Below is an example:

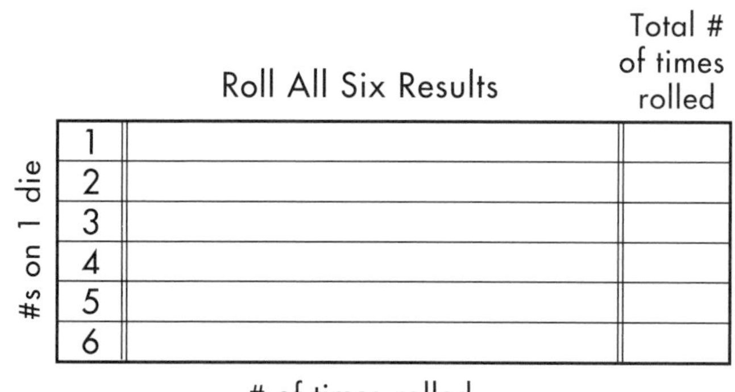

# of times rolled

■ Introduction to a Die: For Both Experiments

1. Ask your students if they have ever used dice before. Have them do a "quick write" in their journals of what they know about dice and how dice are used.

2. Afterward, as students share what they know, record their prior knowledge about dice.

3. Build on the information that they have provided to define a standard one-through-six die.

4. Ask the students if they think a die is a "fair" tool, like a penny, to determine something. Listen to their responses and use their knowledge and any misconceptions to help guide you through the lesson.

   Some students came up with ways to make the die a tool for people to decide something. Here are some ideas:

   - With two people, one person could have all the odd numbers and the other person the even numbers; or one person could have the numbers one through three and the other person, four through six.

   - For four people, each person could have one number and the other two numbers could mean "skip a turn" or "take another turn."

5. Tell students they are going to use a die to conduct an experiment. Continue with Experiment 1 (below) OR Experiment 2 (page 70).

This is an ideal time to review the geometry of the cube, also known as a hexahedron, as a three-dimensional shape with six square faces and eight vertices. Establish that on a standard die, the numbers one through six are represented once each on the six faces.

■ Experiment 1: Horse Race

1. Put the Horse Race transparency on the overhead. Tell students that six horses are in a race. Point out the starting line and the "winner's circle" where a horse wins the race.

2. Explain that in this race a roll of the die will move the horses. After the die is rolled, the horse with the number that was rolled will move one space forward.

3. Put six markers on the board to represent the horses. Have students predict which horse will win and share their predictions with a partner.

4. Give a student a die (the giant-sized die or standard die) and have her roll. Read the number and ask the class which horse will move. Then move that horse one space.

5. Pass the die to another student. Have him roll, and then move the horse that corresponds to his number one space.

6. As the race continues, ask students if they want to change their predictions and why. Play until one horse makes it to the winner's circle.

7. Tell students that they will work in pairs and race their horses. Emphasize that the races need to be played fairly and that students need to keep track of their winners.

8. Distribute materials and have students begin. As they play, post the race results charts. Place a pen near the chart to record results.

9. Circulate as students play. When it appears most students have completed two games, have them freeze their horses and put their dice aside.

■ Collecting and Discussing Race Results

Encourage factual reporting of data, such as Horse #3 won 6 times. Horse #5 won more times than Horse #4.

1. Draw students' attention to the chart. Record the result of the race modeled. Ask a student to report the winners of his races and put tally marks next to the winning horses. Continue having students record until half the class has reported. At this point, ask what information the data provides.

2. Have the remaining students post their data until all of the race results are recorded. Determine how many races each horse won and how many total races were conducted. Guide students into making comparisons about the relative number of wins for each horse.

3. Ask if they think the race is fair to each horse and why. Use probability vocabulary, such as "Does each horse have an equally likely chance of winning the race? Why or why not?" Have students discuss this with a partner.

   Depending upon the outcome of the races, the results may appear fair or not fair! The die is a fair random number generator; however, without enough games it may appear one or two horses are winning a disproportionate number of times. See Background for Teachers in the back of this guide (page 115).

4. Listen to their ideas. Tell the students that they are going to gather more data. Each pair will play as many games as they can until you call time.

5. Circulate as students play. After five minutes, call time and focus the groups again to record the new results. Before recording, ask students to predict how the data will change with additional results.

6. Using a pen in a *different* color, have students record the wins for their horses. Observe if and how the data shifts as more data is added.

7. Ask for their observations with the new results. Does the added data change their thinking about the fairness of a single die? Why or why not?

8. Time permitting, have students play another game. When one horse wins, have them leave all the other horses on the board in their ending positions. Have students look at the board and them ask questions about the race, such as:

   Did all the horses move from the starting position?
   Did any horses stay close together during the race?
   If so, which ones?
   Did you see a pattern of movement in the races?
   Is this a fair race? Why or why not?

■ Theoretical Probability of a Die

1. Depending upon your students experience and abilities, continue the discussion using the theoretical probability of rolling any one number on a standard die. Have students recall that a die has six square faces. On each face there is one number and there is a different number on each face.

2. Use the chances of rolling a two, to illustrate as follows:

$$\frac{\text{\# of ways to get a 2 with one die}}{\text{Total \# of outcomes rolling a standard die}} = \frac{1}{6}$$

3. Continue with the chance of rolling the other numbers. In every case, there is one chance out of a possible six. Because each number has the same probability, it is fair. Therefore, in Horse Race, each horse had an equally likely chance of winning and the race is fair.

4. Remind students about the Law of Large Numbers. According to this law, the more times you conduct a probability experiment, the closer you are likely to get to the theoretical results.

■ Experiment 2: Roll ALL Six

1. Tell students they will conduct an experiment to help determine if a die is a fair or unfair probability tool. The experiment involves rolling a die until each number on the die is rolled one time.

2. Have students talk to a partner to predict how many rolls it will take to roll all six numbers at least one time. Let them share their estimates and determine the highest and lowest estimates they have.

Roll All Six Results

#s on 1 die	# of times rolled
1	
2	
3	
4	
5	
6	

3. Model how to "Roll ALL Six" with a student partner. Have the student be the roller (number generator) and you be the recorder. This allows you to demonstrate how to record. On your board or overhead, create a recording chart. (see sidebar)

4. Have your partner roll the die, and use tally marks to record the number of times each number is rolled. Stop when all six numbers have at least one tally mark.

With the class's help, total the number of rolls for each individual number and calculate the grand total number of all rolls. Here is an example of results:

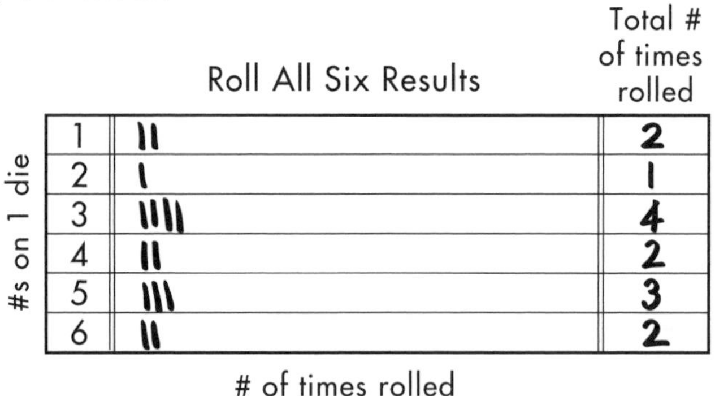

Roll All Six Results

#s on 1 die	# of times rolled	Total # of times rolled
1	II	2
2	I	1
3	IIII	4
4	II	2
5	III	3
6	II	2

5. Compare the actual number of rolls to the high and low estimates. Ask questions such as:

Is the actual number closer to the high or low estimate?
What is the lowest number of rolls needed to roll all six?
What might be the highest number of rolls needed to roll all six?
Is it likely to take more than 50 rolls to roll all six? Why?

6. Review the steps for partners to play:

   a. Decide who will roll first by a roll of the die. Lowest (or highest) number rolls first and the partner records first.

   b. Both students estimate the number of rolls it will take to roll all six.

   c. Using your chart as a model, recorder sets up a chart in her journal to record the data on the number of rolls.

   d. Roller proceeds to roll, while recorder puts a tally mark next to each number rolled. When all six numbers have at least one tally mark, stop!

   e. Count the number of rolls. How did the actual number compare to their estimates?

   f. Switch roles for rolling and recording. Before beginning, estimate again!

   g. Each person rolls two times and records two times.

7. Circulate as students estimate, roll, and record. Listen to their conversations to help you informally assess their skills and abilities. While the students are working, put up the pre-made chart to record their data.

■ Recording the Data Generated

1. When most pairs of students have completed Roll ALL Six four times total, focus the class on their next task. Have the students look at their data and ask questions, such as:

   Did anyone roll all six numbers in just six rolls?
   Who thinks they had to roll the most times to get all six numbers?
   How many people took between 10 and 15 rolls to get all six?

2. Tell students that you want to record the total number of rolls for each number. Point to the chart. First each partner sums their individual tallies for each number, then they add their sums together to get a combined total. For example, the number one was rolled 14 times by Thao and Cody as follows:

   Thao recorded 5 and 6 for a total of 11.
   Cody recorded 2 and 1 for a total of 3.
   Their combined total of number ones rolled was 11 + 3 or 14.

By combining the totals for the four Roll ALL Six experiments they conducted, students have more data than a single experiment to see how the number of times rolled begins to "even out."

3. Have partners calculate their totals for each number on the die. Circulate and assist students as needed.

4. Ask students to look at their totals for each number. Compare the totals to one experiment. Are the total rolls for each number in one experiment relatively equal? Are the total rolls for each number for all four experiments relatively equal?

5. Tell students you want to gather all the data on how many times each number was rolled in this experiment by the class. Have the students look at the total number of ones they rolled. Have each pair report that number and you record it on the chart. Continue with two and so on. Have one partner report for even numbers and the other for odd numbers.

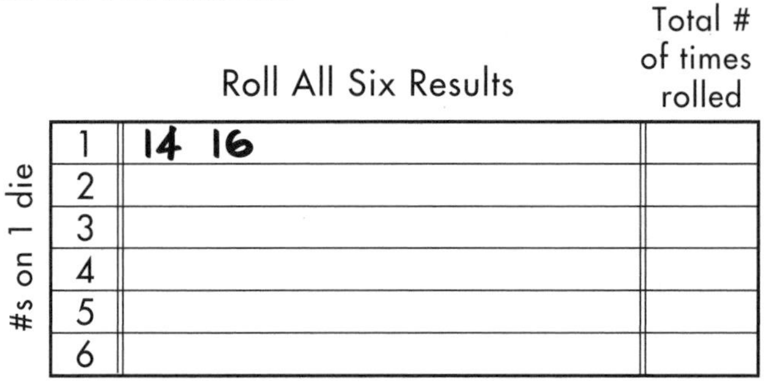

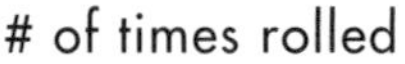

6. Once the data for all six numbers is recorded, determine the class total for the each number. Start with the data for the number one. As a class, calculate the total number of ones rolled by reading the numbers horizontally and summing as you go along. Record the total for the number one.

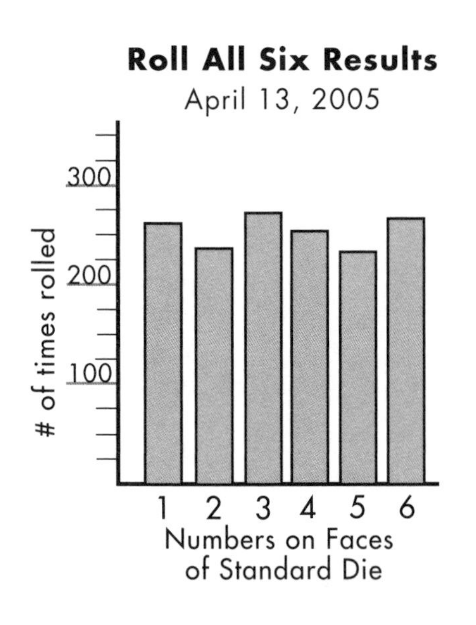

7. Continue with the number two. When you add the results for the number two, have students suggest strategies for adding the numbers. Continue with the remaining numbers.

8. When all the totals are recorded, ask the students how to make a graph for the data. Create a class bar graph. Using their lead, determine a scale for the graph and labels. (See sample on left.)

9. When the graph is completed, ask if it provides information to help determine if rolling one die is fair for all numbers. *Do all the numbers on a die have an equally likely chance of being rolled?*

10. Have students discuss this with their partners or in small groups before listening to their responses.

Depending upon the number of students in the class, the results on the graph may or may not clearly point toward each number appearing a similar number of times. If the results are skewed, continue to collect data until the number of rolls for each number on the die evens out.

11. At the end of the discussion, be sure the students know that the die is a fair probability tool because each number has an equally likely chance of being rolled. There are six faces and each face has a number from one through six; therefore each number has one chance out of six possible chances of being rolled.

■ Theoretical Probability of a Die

1. Continue the discussion using the theoretical probability of rolling any one number on a standard die. Have students recall that a die has 6 square faces. On each face there is one number and there is a different number on each face.

2. Use the chances of rolling a two to illustrate as follows:

$$\frac{\text{\# of ways to get a 2 with one die}}{\text{Total \# of outcomes rolling a single die}} = \frac{1}{6}$$

3. Continue with the chance of rolling the other numbers. In every case, it is one chance out of a possible six. Because each number has the same chance, it is fair. However, it is not likely that you will roll the numbers one through six in just six rolls! Each time the die is rolled, there is a one-in-six chance for each number to be rolled.

4. Remind students about the Law of Large Numbers. According to this law, the more times you conduct a probability experiment, the closer you are likely to get to the theoretical results.

Session 2: Off to the Races!

■ What You Need

Materials:

- ❑ cardstock
- ❑ paper
- ❑ overhead transparencies
- ❑ overhead pens
- ❑ 12 markers (beans, cm cubes, plastic circles)
- ❑ 2 dice
- ❑ chart paper with 1-inch grids
- ❑ pens, 2 different dark colors

For each pair of students:

- ❑ 2 dice
- ❑ 1 Double Dice Derby board
- ❑ 12 markers (beans, cm cubes, plastic circles)

For each student:

- ❑ math journal

Optional:

- ❑ 1 giant-sized pair of standard dice

■ Getting Ready

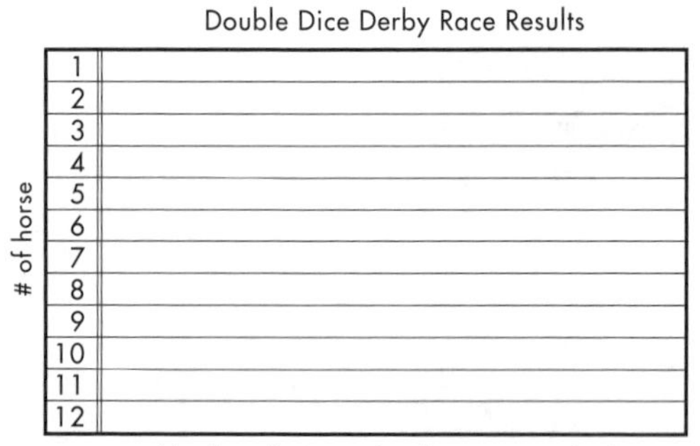

1. Using the Double Dice Derby board with 12 horses (page 86),
   a. Duplicate one cardstock copy per pair of students.
   b. Duplicate one paper copy for each student for the homework assignment.
   c. Duplicate a copy on an overhead transparency for demonstration.

2. Gather 12 markers for each pair of students and the same for yourself to introduce the game. The markers represent the 12 horses in the race. Check that the markers fit on the board.

3. Gather two dice, each a different color, for each pair of students. Gather dice, either large or small, for yourself to introduce the game.

4. Use a sheet of chart paper, approximately 24 by 36 inches, to make a chart to keep track of the winners of the Double Dice Derby. List the numbers of the horses vertically along the left side of the paper and label the chart as in sidebar.

■ The Race Is ON!

1. Put the Double Dice Derby transparency on the overhead. Tell your students they will race horses in the Double Dice Derby. Each of the 12 horses is competing to be the first to enter the winner's circle.

2. As the name says, the horses will move forward by the roll of **two** dice. Tell them that, in this race, the **sum** of the dice indicates which horse can move forward one space. Ask them to predict which horse they think will win the race.

3. Set up the Double Dice Derby overhead with 12 markers to represent horses. Enact the race with the class. Have one student roll the dice and tell the two numbers. Then the class adds the numbers and the sum determines which horse leaves the "Starting Gate." Move that horse one space.

4. Pass the dice to another student to roll. Students add the numbers rolled, then move the horse that corresponds to the sum. Continue until one horse wins OR until your class understands how to play. Be sure to emphasize that once one horse wins, the race is over—no second and third place winners!

5. Draw students' attention to the chart where they will keep track of the winners of each race. As soon as a horse wins, one person from the pair will come up to the chart and record the winner with a tally mark. Remind them that after four tally marks, a slash indicates five. Have only one color pen by the chart to record race results.

6. Distribute materials to partners and have them begin. Be aware that the races get very lively. At times, it is important to remind them that the horses are competing—the students are not competing with one another! Have each pair play as many games as they can until you call time.

7. Circulate as they play. Keep a watchful eye on the data on the chart. Let students continue to play until there is sufficient evidence to see that some horses (most likely 6, 7, and 8) are winning more frequently than others.

■ Discuss Results

1. When you call time, have students move their horses back to the starting line and set their dice on the board. Focus their attention on the Double Dice Derby Results chart. Have them talk to their partners about the results.

2. Depending upon your students' responses, you will be able to build on what they know. Here are some student responses and possible teacher considerations:

   "Number One is never going to win." Or, *"It's not fair to Number One!"* Be sure to pursue this! Ask students to explain why.

   "Number Seven wins a lot because seven is a lucky number!"

   Though the observation that 7 won "a lot" is accurate, the rationale is not accurate. However, it is important to allow students to conjecture and check to see if others agree with this type of thinking. The conjecture does not exist as fact! With more evidence and investigation in Session 3, students will see the reason why 7 wins "a lot."

   "Big numbers don't come up a lot."

   Ask which numbers. This student elaborated by saying that you can only get a 12 with a six and a six and an 11 with a five and a six.

   "Horse Number Two hardly moved."

   Encourage the student to explain why that horse did not move often.

   Sometimes students notice the frequency of the numbers in the center of the data. For example, one student said, *"Six and seven won a lot of times."* Another student chimed in, *"There are lots of combinations for six and seven."* This can spark recording the combinations to make six and seven.

   After the combinations were recorded for six, one student piped up and said, *"I see the commutative property. One plus five and five plus one."*

   Be sure to build on these teachable moments. Have students take their dice and create both 1 + 5 and 5 + 1. The result is six regardless of the color or order; however, there are two distinct ways to get a six with a five and a one.

Ways to Make Seven with Two Dice

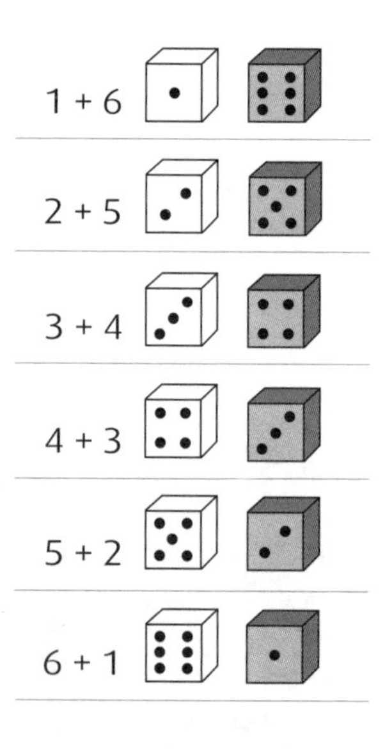

3. If your students are not forthcoming with observations, you may need to pose additional questions to scaffold their learning, such as:

   Which horse(s) won most frequently?
   Which horse(s) never won?
   Did any horse(s) never leave the starting gate?
   Why do you think some horses are winning more/less than other horses?

4. Record the combinations for the horse numbers. For example, for Horse #4, generate the three different combinations that sum to four using two dice.

Horse #4	
1	3
3	1
2	2

   For some students, it will be helpful to actually manipulate the dice so they can clearly see the difference between the combinations 1, 3 and 3, 1.

5. For **Homework,** have the students play the Double Dice Derby game three more times with a family member or friend. Brainstorm items they can use as markers for the horses. Distribute game boards and be prepared to lend dice overnight for those students who do not have any at home. Have students record the results of the three races in their math journals.

Session 3: Keeping Track

■ What You Need

Materials:

- ❑ overhead transparencies
- ❑ overhead pens
- ❑ pens, 2 different dark colors
- ❑ chart paper or sentence strips

For each student:

- ❑ 1 copy of Keeping Track data sheet
- ❑ math journal

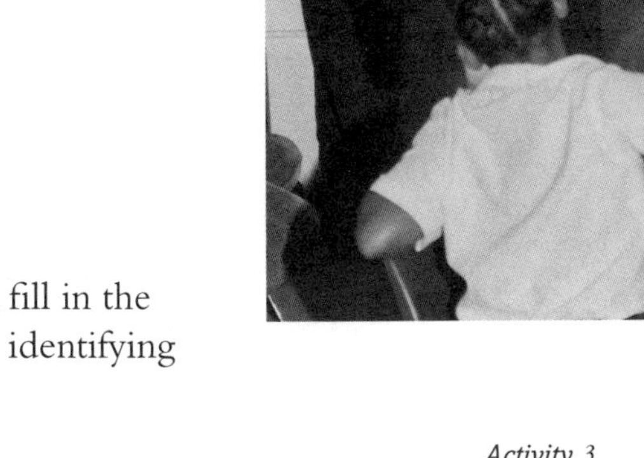

■ Getting Ready

1. Using the "Keeping Track" chart on page 87,
   a. Duplicate one copy of the chart for each student.
   b. Make two overhead transparencies. On one overhead, fill in the chart with black permanent pen. This will be used for identifying

patterns with your students. The blank one will be used to introduce the activity.

2. Gather a pen to record the race results from homework. Select a pen in a *different color* from the one used by students for the initial race.

3. For Grades 4 and 5, pre-make a number line from 0–1 that is 38 inches long using chart paper or sentence strips. Situate the zero 36 inches from the number one, so that there are 36 1-inch increments along the line. These will be used to identify the theoretical probability for the sums of two dice.

■ More Race Results

1. Have students share the results of the three races they conducted for homework. Record the wins for each student on the Double Dice Derby Results chart. *Be sure to use a different colored pen so that the effect of adding new data to the chart will be very visible.*

2. After their data is added, ask the students to make observations about how the data has changed. Listen to their ideas. Ask students if the additional data can help them determine if the race is fair and if there is (are) a horse(s) more likely to win the race.

3. Build on their responses to delve further into the reasons behind the race results. The following are some examples from students during this discussion:

   "Horse Seven is lucky."
   Often students hold onto the idea that the number 7 is lucky. Remind the students how the horses move! Build on students' ideas that lead to the combinations on the dice that add up to 7.

   "The horses in the middle won more times."
   Ask which horses they mean and why they think those particular horses won more times. This can lead to a comparison of the combinations that allow the middle horses to move.

   "The race is not really fair for Horse 2 and Horse 12."
   Ask students to explain why. Again, look at the combinations that will allow Horses 2 and 12 to move. In each case, only one combination is possible for each of these two horses.

   "It evened out for some of the horses. Number Six won about the same number of times as Number Eight, and Number Four and Number Ten are also close."

4. Often a student observes the following: "If you turn the chart, it looks like a mountain. It goes up and then down." With or without this lead, take the opportunity to create a line graph of the race results from the class graph:

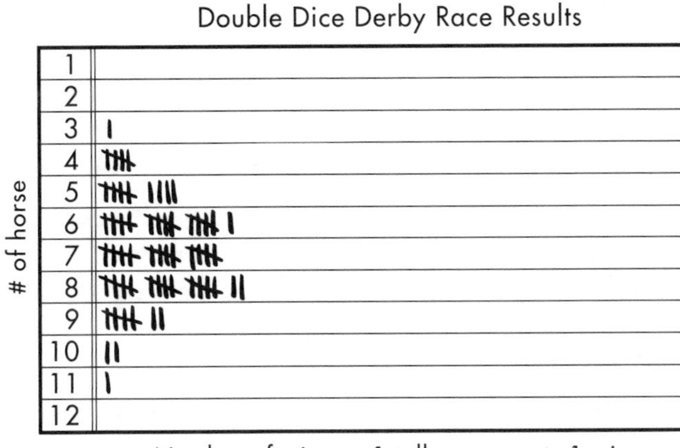

Create the structure of the graph and have students help you label it. Then add the results to the graph using a point to indicate the number of wins for each horse.

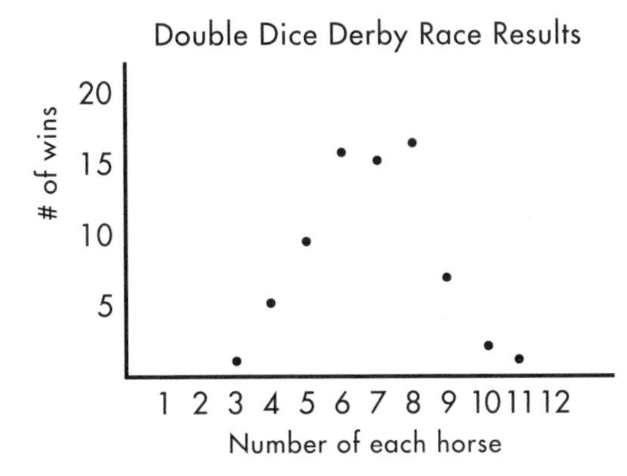

Tell students that to look at the trend of wins, you can connect the dots and do so. This is a line graph. Ask what the shape reminds them of.

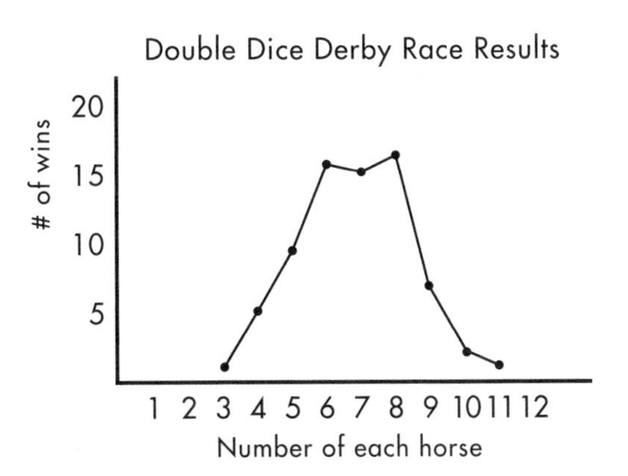

Note how the race results parallel the theoretical.

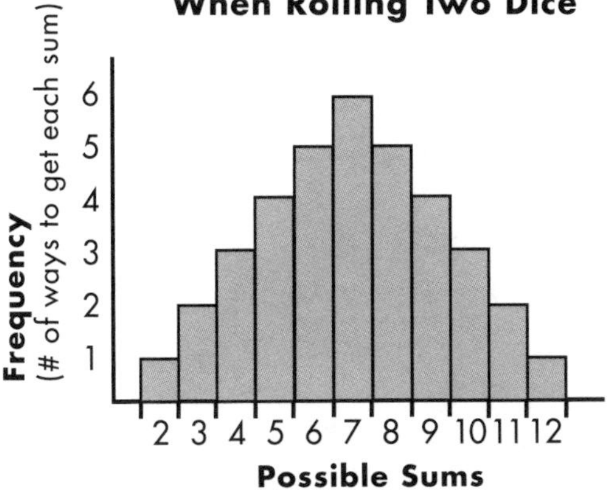

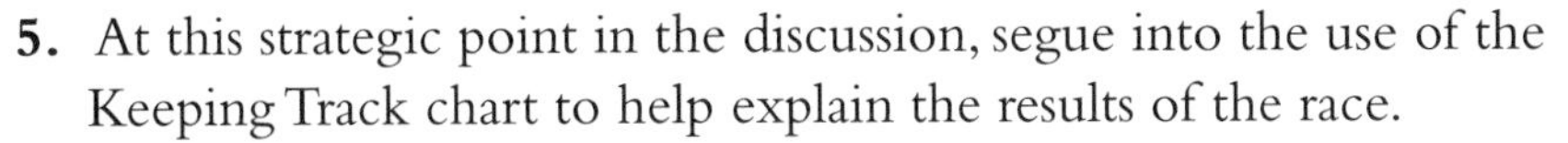
5. At this strategic point in the discussion, segue into the use of the Keeping Track chart to help explain the results of the race.

■ Keeping Track

1. Put the blank Keeping Track chart on the overhead. Tell students that this chart is a tool to show all the possible combinations for the sums of two standard dice.

2. Point out the numbers 1 through 6 along the top horizontal line. These numbers represent the possible outcomes when one die is rolled. Then look along the vertical line on the left side of the chart. The outcomes for one die are also recorded there.

3. This Keeping Track chart will show all the possible combinations when two dice are rolled together. Since in this experiment the outcome of the dice rolls were added, draw an addition symbol at the top left corner of the chart. This defines the operation being used to fill in the chart.

4. Have students help you fill in the first few numbers in the top row. Once they understand how to complete the chart, tell them that each person will get a chart to complete. When they finish filling in the sums, they will look for patterns on the chart.

5. Distribute the Keeping Track charts. As students work, have your completed Keeping Track overhead readily available with colored pens.

6. Circulate as they work and assist students as necessary. Encourage students to look for as many patterns as possible.

7. When all students have completed the chart, focus their attention on your completed chart. Be sure that their charts match yours. Ask for the patterns that they found on the chart. Use a different color marker to highlight each pattern shared by students.

Here are some of the patterns that a class of fourth-grade students identified:

"Every time you move from the left to the right on a line (row) the number gets bigger by one." Identify this is as a "+1" pattern.

"I see a 2, 4, 6, 8, 10, 12 pattern if you go on a diagonal down from the top corner to the bottom. It is a plus two pattern."
Have students check all the diagonals starting on the top left and going down to the bottom right. They all have a "+2" pattern.

"If you go on a diagonal from the top right and down, you go by zero. The numbers are the same. Except the two and the twelve only show up once."

"If you start at the two (top left corner) and go across and then when you get to the seven you go straight down, you have the numbers 2 to 12 which are all the numbers you can get when you roll two dice."

"If you go straight down (vertically) you also count by ones."

"You go backwards by one if you go from right to left or from the bottom to the top. It is a '−1' pattern."

"When you add doubles the sum is always an even number. Even when the two numbers are odd."
This is an opportunity to review the addition of even and odd numbers, time permitting.

"The seven appears six times and that's how many ways we've got to roll seven."

"The seven runs right down the center of the chart. And six and eight are next to it on each side and they have five combinations. Then comes five and nine, which have four combinations."

8. At the appropriate point in the classroom discussion, ask students how many combinations are possible when you roll two dice. Have them talk to a partner or in small groups before sharing ideas.

9. Listen to their answers and the reasoning behind them. Some students may add up all of the combinations they generated for each sum (2-12), while some may use the nature of the six-by-six chart to multiply and get 36. One student said that each die had six faces, so she multiplied six-by-six to get 36.

Some students may not make the connection between the chart and the number of possible combinations. Be sure that you guide them to understand.

10. Connect the Keeping Track chart to the line graph. Ask how the two are related. Remind the students that the graph represents the outcomes in an *experiment,* while the chart shows *in theory* the outcomes in rolling two dice.

11. Provide a writing prompt to assess students' understanding. One is provided at the end of this session, page 84.

■ More on the Theoretical (For Grades 4 and 5)

1. Guide students in expressing the probability of each outcome as a fraction with the Keeping Track chart as a tool.

2. Reestablish the total number of outcomes possible when rolling two dice to equal 36.

3. Ask how many ways there are to get a 7 with two dice. If they do not remember, have them look at their Keeping Track chart and count the number of sevens. There are 6 sevens out of the 36 total outcomes. Ask how to express that as a fraction. Guide students if they do not remember.

Some students may immediately know an equivalent fraction for $6/36$ is $1/6$. Agree that $6/36$ and $1/6$ are equivalent fractions, though do not feel compelled to prove this to the entire class. The goal here is the relative size of the fractions for each outcome and what that represents about the likelihood of a number being rolled.

$$\frac{\text{6 ways to get a seven}}{\text{36 total outcomes}} = \frac{6}{36}$$

4. Next, look at the number of ways to get a 12. There is only one way. Ask how to express this. ($1/36$) Ask if there is any other number that there is only one way to roll. The number 2 only has one possibility out of 36 total outcomes.

$$\frac{\text{1 way to get a 12}}{\text{36 total outcomes}} \text{ and } \frac{\text{1 way to get a 2}}{\text{36 total outcomes}} = \frac{1}{36}$$

5. Continue with the remaining numbers. For the numbers 3 and 11, there are 2 possible ways to roll out of 36 total outcomes, $2/36$.

   For the numbers 4 and 10, there are 3 possible, $3/36$.
   For the numbers 5 and 9, there are 4 possible, $4/36$.
   For the numbers 6 and 8, there are 5 possible, $5/36$.

6. Have students count how many times a 1 appears on the Keeping Track chart. It can never be rolled when you are adding two standard dice. Therefore, there are no possible ways out of the 36 total outcomes. Express this as the fraction $0/36$ or 0. This is another example of an event that will never happen.

7. Put up the new number line from 0 to 1. Ask students what the number 0 represents when talking about the chances of an event happening. (never will happen). What does 1 represent? (always will happen) Label those points.

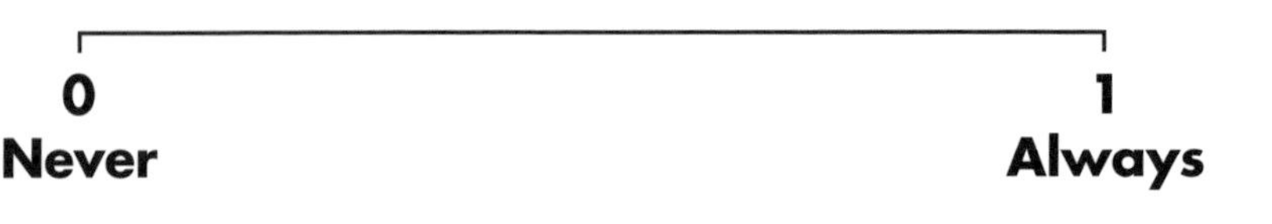

8. Label the number line with the title Probability for the Sums for Two Dice. Use this line to situate the probability for each of the numbers on the number line. Since there are 36 possible outcomes, mark the 36 equal increments on the number line.

Probability for the Sums for Two Dice

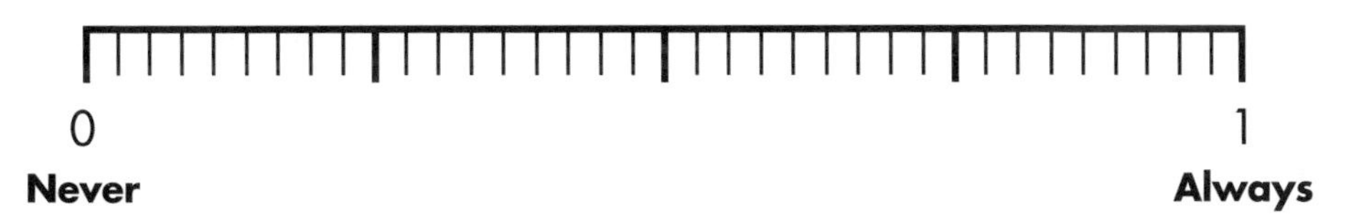

9. Ask where the probability of rolling a 7 with two dice would be situated. Have a student go to the number line and locate the point. Check with the class for agreement. Record on the number line as follows:

Probability for the Sums for Two Dice

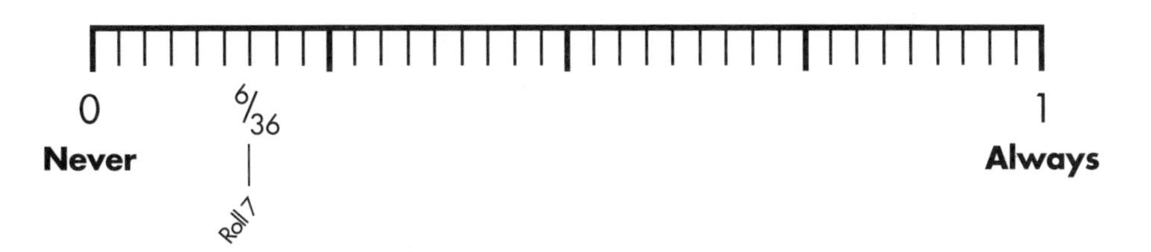

10. Continue with the other sums and their probabilities. The number line will look like this.

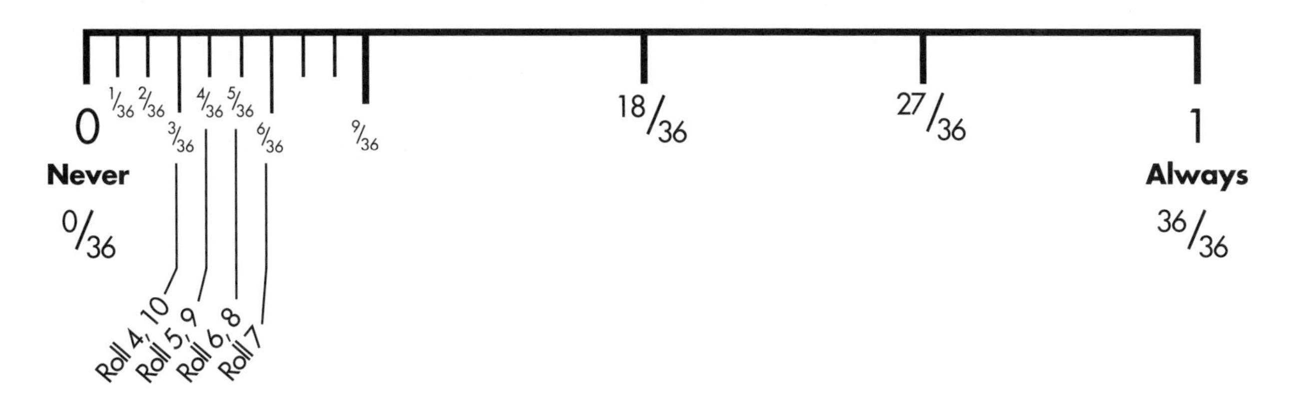

11. This shows the theoretical probability of each sum of the dice. However, when conducting experiments, such as the Double Dice Derby, the outcome of rolls does not always match the theoretical!

■ Wrapping Up Through Writing
(For all students)

Provide a writing prompt to assess your students' understanding of the activities, such as:

Mugsy is going to play the Double Dice Derby game with his friends. He wants your advice on what horse he should pick to win. Tell Mugsy what horse you think he should pick.
Be sure to explain WHY he should pick that horse.
Use all the information you know about the race to help with the explanation. You can use graphs and charts.

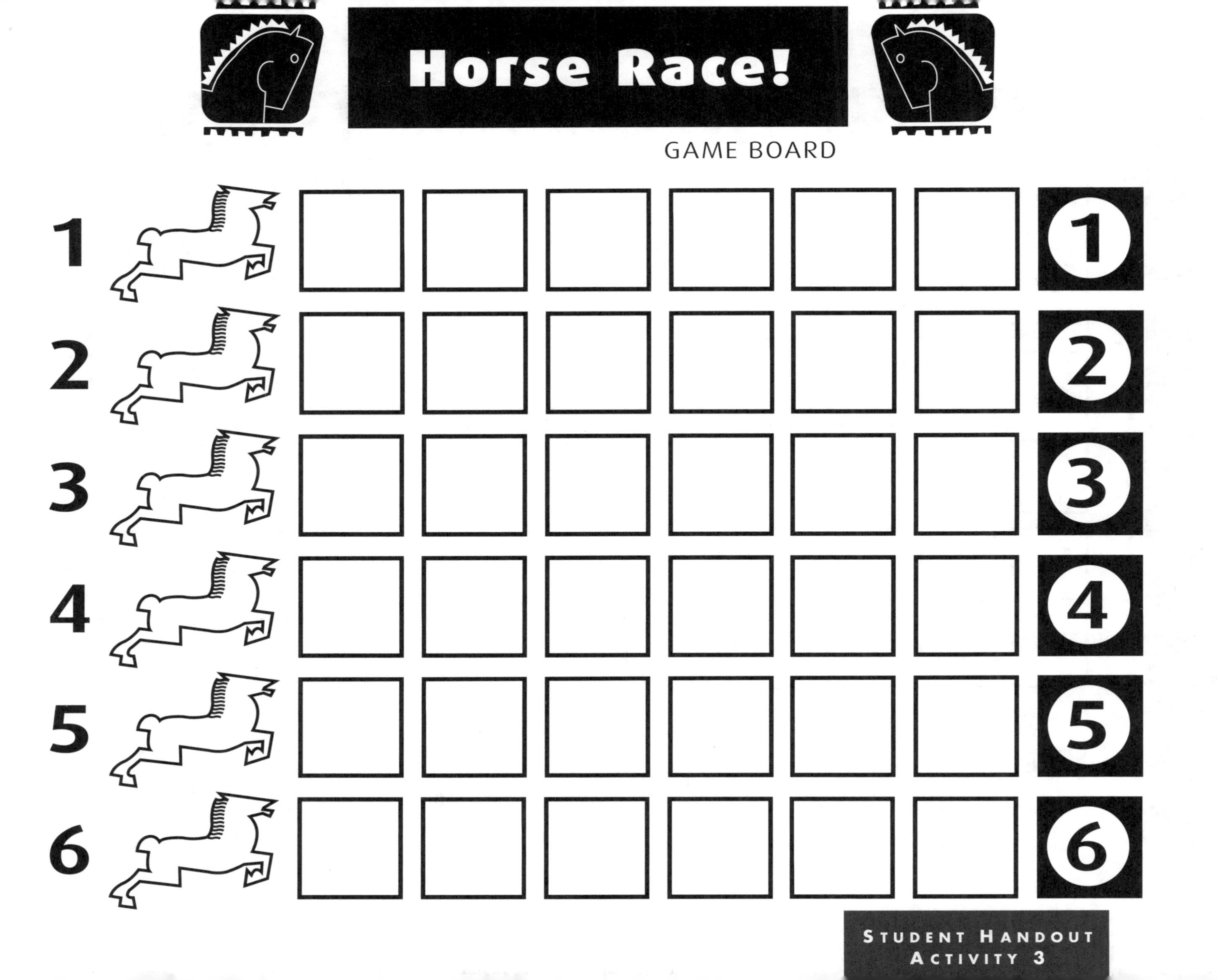
Horse Race!
GAME BOARD
1
2
3
4
5
6
STUDENT HANDOUT
ACTIVITY 3

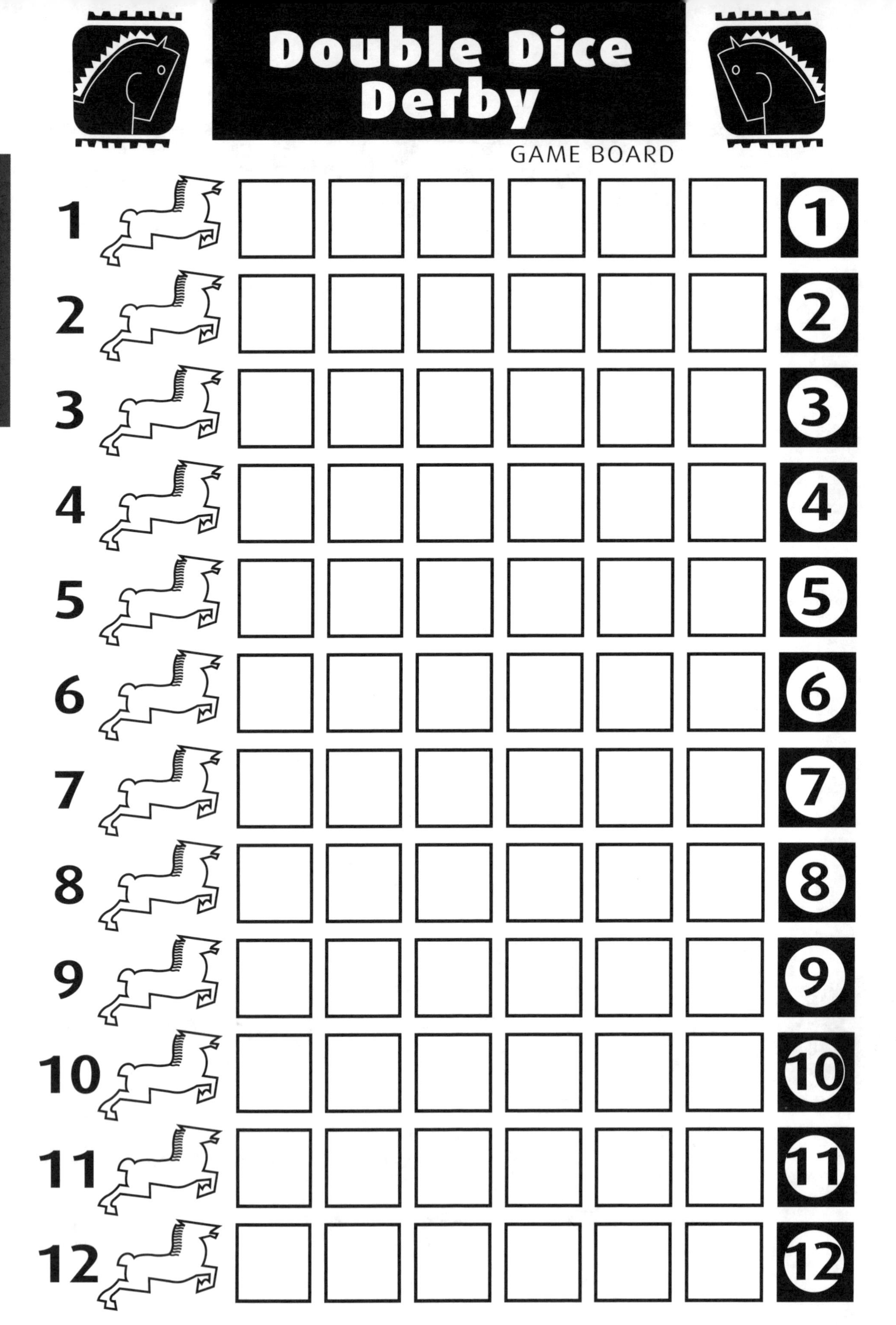
Double Dice
Derby
GAME BOARD
1
2
3
4
5
6
7
8
9
10
11
12
1
2
3
4
5
6
7
8
9
10
11
12

Keeping Track

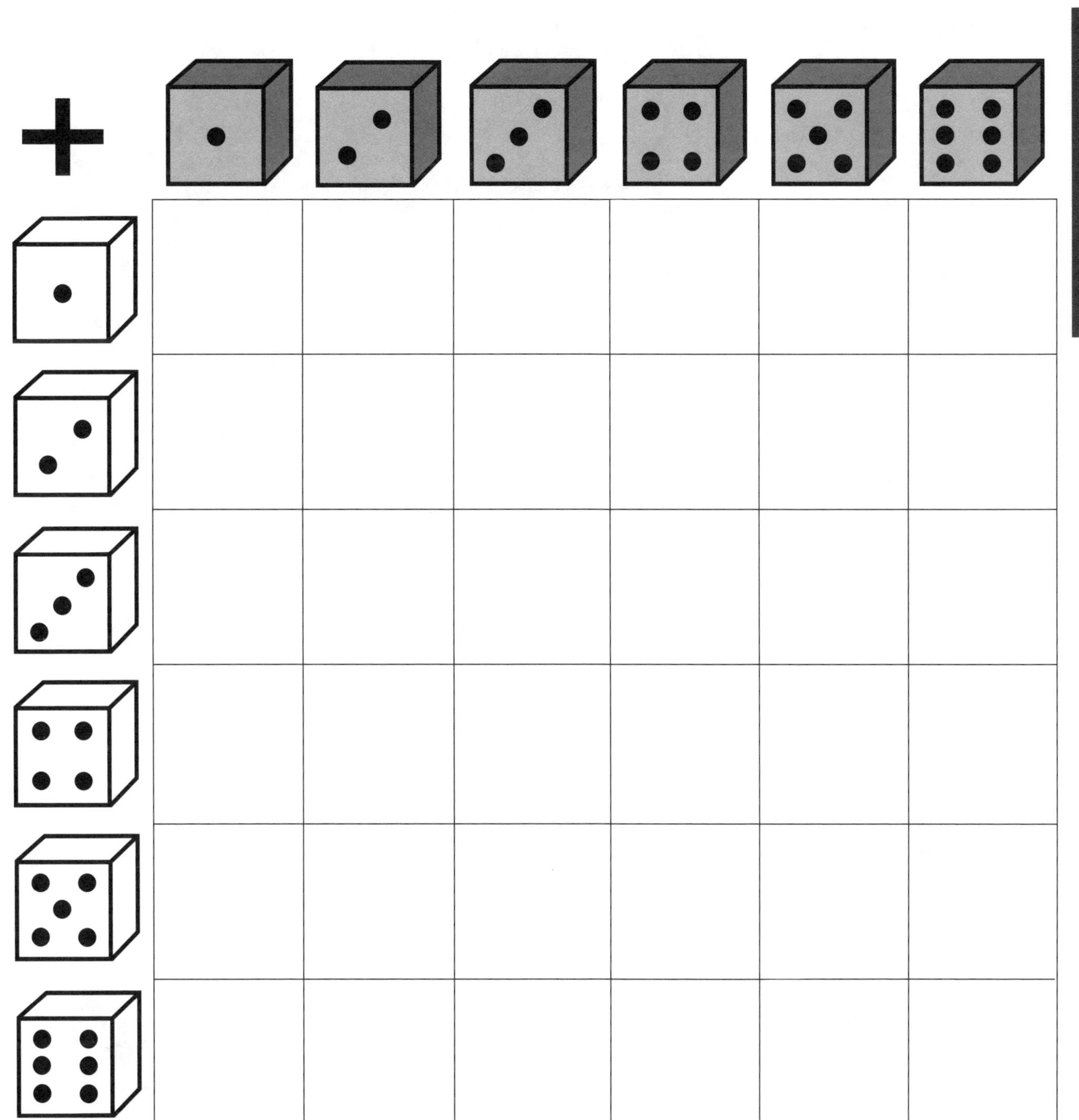

Overview

Game Sticks is based on one of many Native American games of chance. In this version, four two-sided sticks are used to play the game. Students create a set of sticks and then play the game with classmates. After students have experience with the game, they gather data about the outcomes of the tosses to investigate the probability behind them.

To launch this activity, students learn that Native Americans played games of chance and are introduced to Game Sticks. In California, sticks were made from thin tree branches, such as alder, and had carved designs on one side. Students will create their sticks from wooden tongue depressors. After seeing models of sticks, students are given a set of four plain wood sticks to design their own set of game sticks for homework.

In Session 1, the game is introduced with the class divided into two groups. The object of the game is to obtain a set of 10 counters that are awarded based on the outcome of tosses. The procedure of tossing, reading the outcome of design and plain sides, and collecting counters (points) is modeled. Depending upon the combination of design sides and plain sides, a player takes between zero and three counters.

Given the probability and how points are assigned to the outcomes, it is likely that every student will have a chance to toss the sticks once in the class game. After students understand the game, they play in partners and have first-hand experience with the outcomes. For homework, they play the game at least two additional times, with a focus on the outcomes.

Session 2 begins with students sharing their experiences playing the game at home. During a class discussion, the possible outcomes are generated and recorded. Students make conjectures about which outcome(s) occurs most frequently in the game. To test their ideas, the students make data sheets in their journals to keep track of 10 tosses. After partners compare their results, the class results are compiled to provide more data to analyze the game. By carefully examining the different ways that each outcome can be generated with the four sticks, the students come to understand the frequency of the outcomes. Students in fourth and fifth grade determine the theoretical probability for each toss as well. The session ends with a writing prompt to assess student understanding.

Where's the Math?
Game Sticks is related to the first probability experiment that students performed with pennies in this unit. Each game stick has two possible outcomes, just as a penny did. However, in this case four sticks are tossed and there are five different outcomes that can occur based on the combination of plain and design sides. Further, there are 16 different ways to arrange those five outcomes. Students first explore the probability by playing the game and observing outcomes. Building on this knowledge, they collect and analyze data from an experiment focused on outcomes. Finally, they determine the theoretical probability for each outcome to further understand the underpinnings of the game.

■ What You Need

For the class:

- ❑ 8 tongue depressors or craft sticks 3/4-inch wide by 6 inches long
- ❑ 10 counters
- ❑ colored markers
- ❑ chart paper

For the each student:

- ❑ 4 tongue depressors or craft sticks 3/4-inch wide by 6 inches long
- ❑ 10 counters, plus some for homework
- ❑ math journal

■ Getting Ready

1. Read over the information on "Native American Game Sticks," pages 118–122.

2. Gather either tongue depressors or craft sticks, so that each student will have four, and so that there are a few extras for any "mess ups." You will need eight for class activities. The sticks are commonly available from teacher supply, craft, and drug stores. These will be distributed for students to design, as homework before the first session.

3. Create eight game sticks by drawing designs with colored markers on one side of each stick. Write your initials on the opposite side. When you assign the homework assignment for students to make their sticks before Session 1, you will show these. They will also be used in Sessions 1 and 2.

4. Play the game a few times with a partner to learn the game, get a feel for the frequency of the outcomes, and gain a time frame for playing.

5. Gather counters such as toothpicks, cubes, or beans as follows:
   - 10 counters to model the game
   - 10 counters per pair of students
   - additional sets of 10 counters for students who may need counters for homework

6. On chart paper, record the number of counters or points for the outcomes as follows:

Outcome	**Points**
All Design	3 Counters
All Plain	2 Counters
2 Design/2 Plain	1 Counter
Other Tosses	0 Counters

   Include an illustration next to each outcome.

7. Read Session 2. If the session has too much to cover in your math class period, end the session after the experiment one day and analyze the game on another day.

■ Introducing Game Sticks

1. Tell the students that they will be learning a new game based on Native American gambling games. The game is called Game Sticks and it requires four sticks that are decorated on only one side.

2. Show a few of the sticks that you have made. Tell the students that Native Americans usually made these sticks from thin tree branches that were split in half. Then they decorated one side with designs, often by burning.

3. For **homework,** have the students decorate a set of four sticks on one side. On the reverse side, have them write their initials. Tell them that when they return to school with their decorated sticks, they will learn the game.

You may want to show students the Traditional Designs sheet on page 101 to spark design ideas.

Session 1: Game Sticks in Action

1. Tell the students they will learn how to play Game Sticks as a whole class using your sticks. Let them know that when Native Americans played the game, they brought objects to gamble, but the students will not be gambling for objects at school.

2. Explain that the object of Game Sticks is to obtain all 10 counters. Post the chart with the outcomes. Show and explain the outcomes of the sticks and connect them to the related point(s).

Outcome	Points
All Design	3 Counters
All Plain	2 Counters
2 Design/2 Plain	1 Counter
Other Tosses	0 Counters

3. Divide the class into two teams. Put 10 counters in a central location between the teams. Tell the students that the sticks will move from student to student, and everyone will have a chance to toss the sticks for their team. Give one person on each team a set of four game sticks.

4. To find out which team will toss first, have the person on each team with the sticks give them a toss. Explain that whichever team has the most design sides facing up goes first. Determine which team goes first and then begin play.

5. After the first person tosses her team's sticks, count the number of design and plain sides facing up. Check the outcomes chart to see if that toss receives any points. If so, that team takes the appropriate number of counters.

6. Then a player on the opposite team tosses his team's sticks. Check the number of design and plain sides and take counters as indicated by the combination. Teams alternate turns with a different student tossing each time.

7. As students play, it is likely that all the counters will be taken from the central area and neither team will have all the counters. The game still continues until one team obtains all the counters. If a team's toss warrants counters, the opposing team has to give counters from their stash of counters! You may watch the counters move back and forth from team to team for quite a while!

8. Play continues until one team obtains all the counters OR until every person on each team has had a chance to toss the sticks.

9. Review how to play the game and be sure everyone understands the rules.

Native Americans often used songs or chants to encourage their team. Your students may wish to sing or chant encouraging words when it is their team's turn. Make it clear to the students that only positive cheers for their team are acceptable, and that no jeers or other negative comments toward the other team are allowed.

■ Playing the Game with a Partner

1. Have partners play the game until one player has all the counters.

2. As students play, circulate and answer questions.

3. When most students have played the game at least two times, focus the group for a class discussion.

4. Ask for their observations about the game. Ask how many students tossed all design sides. How many tossed all plain sides? What seemed to be the most common toss?

5. Tell the students their **homework** assignment. Explain that they will play the game at least two times with a family member or friend. When they play, they should pay attention to the outcome(s) that occur most frequently and those that occur less frequently.

Session 2: Investigating the Outcomes

1. Start the session by asking students for their observations about the games they played for homework.

2. With student assistance, record all the possible outcomes they observed and draw those outcomes on the board as shown in the illustration on the right.

3. Have the students discuss with a partner which outcome they think is most likely and why.

4 design/0 plain

3 design/1 plain

2 design/2 plain

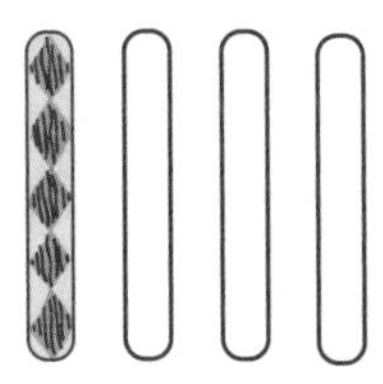

1 design/3 plain

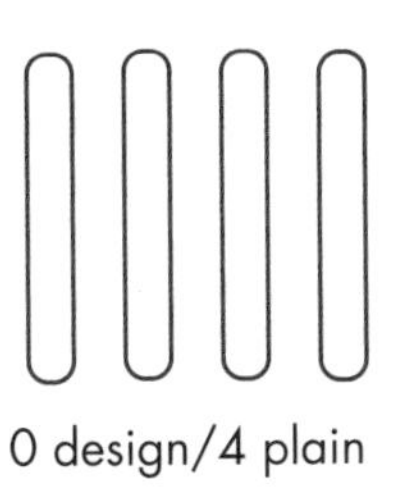

0 design/4 plain

■ Conducting an Experiment

1. Tell the students that they are going to conduct a probability experiment to help determine which outcome is most likely. One partner will toss their sticks 15 times and the other partner will record the outcomes. Then they will switch roles. Finally, the results for the class will be tabulated to help them analyze the game.

2. Model how to conduct an experiment. Select a partner. Create a data sheet on your chalk or wipe-off board. The data sheet will look as follows:

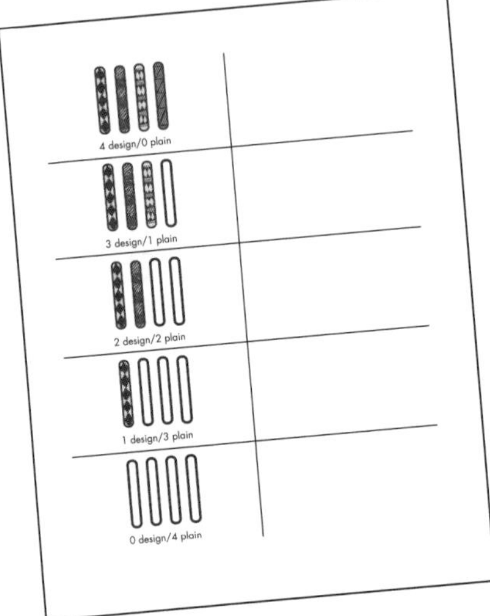

3. Have your partner toss the sticks 15 times and you record each toss with a tally mark. Keep track so that after 15 tosses, the "tosser" stops. Total the number of times each outcome occurred, such as:

4 design/0 plain	I	1
3 design/1 plain		0
2 design/2 plain	~~IIII~~ I	6
1 design/3 plain	~~IIII~~ I	6
0 design/4 plain	II	2
	Total Tosses:	15

4. Tell students that you and your partner would then switch your roles so that each person has an opportunity to toss the sticks 15 times and to record the results of her partner's 15 tosses.

5. Be sure the students are clear on what they are meant to do. Circulate as students conduct the experiment. Assist as needed.

6. After both partners have tossed and tallied, focus your students' attention. Given the results of two experiments, ask students what they think the most common outcome for the sticks is. Listen to their ideas.

7. Tell them that to have more data to consider, you want to collect the results from all of their experiments. To collect the data efficiently, have students combine their results. Do an example on the board for clarity, such as:

	Student 1	Student 2	Total
4 design/0 plain	1	1	2
3 design/1 plain	0	4	4
2 design/2 plain	6	5	11
1 design/3 plain	6	3	9
0 design/4 plain	2	2	4

8. While students are adding their tallies, prepare a chart on the board to record the numbers for each outcome. Here is the start of the data from one class:

Total # of Times Each Outcome Occurred in Class Experiment

4 design/0 plain	2	1	...
3 design/1 plain	4	5	...
2 design/2 plain	11	14	...
1 design/3 plain	9	8	...
0 design/4 plain	4	2	...
	30	30	

9. After all the results for the class are posted on the board, calculate the total number of times each outcome occurred. For example,

Total # of Times Outcome Occurred in Class Experiment

				TOTAL
4 design/0 plain	2	1	...	24
3 design/1 plain	4	5	...	67
2 design/2 plain	11	14	...	88
1 design/3 plain	9	8	...	72
0 design/4 plain	4	2	...	19

10. Ask what new insights the students have with this additional data. Can they say with more conviction that a particular outcome occurs more frequently than the others? Provide time for partners to think and share with one another.

11. In terms of the game, both the 3 design/1 plain and the 1 design/3 plain tosses result in no points being won. Considering those two outcomes as one outcome, the total number of times the two occurred in the experiment is 139. Does that help explain why so many tosses resulted in no points?

■ Analyzing the Results

1. Tell the students that you want to analyze the game another way. They will use their game sticks to better understand the outcomes.

2. Tell them to situate their game sticks so that all design sides face up. Ask how many different ways it is possible to get this outcome. Agree there is only one way to get all designs up. Record that result.

   D D D D

3. Continue with all plain sides up. Have them arrange their sticks that way and agree again that there is only one way to get this outcome. Record that result.

   P P P P

4. Next have students place three design sides and one plain side up. Ask a student to read you the order of their sticks and record it. Check to see how many students have that order. Ask for a different arrangement of three designs and one plain.

5. Guide the students to manipulate their sticks to concretely see that there are four possible ways to get the "three design and one plain" outcome.

   D D D P
   D D P D
   D P D D
   P D D D

Since each stick is operating independently of the other sticks, there are four different ways that each outcome could occur for a three design and one plain toss OR a three plain and one design toss.

6. Continue with the three plain and one design outcome. Again, have students manipulate the sticks to concretely see the outcomes!

   P P P D
   P P D P
   P D P P
   D P P P

7. Ask students to predict how many ways they can create a two design and two plain toss. Then have them manipulate their sticks to see if their predictions were accurate. Have partners generate the list

of different ways to get that toss. As a class generate the six possible ways to get this outcome.

D D P P
P P D D
P D D P
D P P D
P D P D
D P D P

8. Ask students what this information tells them about the likelihood of getting each toss. Have them discuss with a partner, and then as a class. Ask how the outcomes affect the points you receive in the game.

9. During this discussion, be sure students see that they are most likely to get a toss that does not get any points. Though the 3 design/1 plain and 1 design/3 plain outcomes are distinct, in the game the difference does not play a role in determining points. As a result, with four ways to get each outcome, there is a total of eight out of 16 possible. This is more than the six possible ways to get the 2 design/2 plain outcome.

10. For third-grade students, end the session with a writing prompt. Several are suggested at the end of the session. (page 99)

■ Delving Into the Theoretical (for Grades 4 and 5)

1. Have students total all the possible ways for four two-sided sticks to fall. [16]

2. Guide students in determining the theoretical probability, expressed as a fraction, for each outcome.

   a. Start with the toss for ALL DESIGN sides. How many possible ways are there to get that toss? Then express it as a fraction:

$$\frac{\text{Total \# of Ways to Get “All Design” Outcome}}{\text{Total \# of Outcomes for 4 Sticks}} = \frac{1}{16}$$

   b. Continue with ALL PLAIN sides. Have students talk to one another about their ideas on the probability for this occurring. Record it as another fraction.

$$\frac{\text{Total \# of Ways to Get “All Plain” Outcome}}{\text{Total \# of Outcomes for 4 Sticks}} = \frac{1}{16}$$

c. Continue with the THREE DESIGN and ONE PLAIN outcome. Have students talk to one another about their ideas on the probability and how to write it as a fraction.

$$\frac{\text{Total \# of Ways to Get 3 Design \& 1 Plain Outcome}}{\text{Total \# of Outcomes for 4 Sticks}} = \frac{4}{16}$$

Connect this outcome to THREE PLAIN and ONE DESIGN side. Have a student come to the board to record the fraction that represents the theoretical probability.

$$\frac{\text{Total \# of Ways to Get 3 Plain \& 1 Design Outcome}}{\text{Total \# of Outcomes for 4 Sticks}} = \frac{4}{16}$$

d. Conclude with the final possible outcome—TWO DESIGN and TWO PLAIN. Ask students to discuss how to write the theoretical probability. Have a student record it on the board for the class.

$$\frac{\text{Total \# of Ways to Get 2 Design \& 2 Plain Outcome}}{\text{Total \# of Outcomes for 4 Sticks}} = \frac{6}{16}$$

3. Ask students to predict what the sum of all the fractions representing the outcomes will equal and listen to their predictions. Have them determine the sum. ($^1/_{16} + {}^1/_{16} + {}^4/_{16} + {}^4/_{16} + {}^6/_{16} = {}^{16}/_{16} = 1$)

4. Have the students discuss why the sum is one. Have them think back to the zero-to-one scale for the probability of an event. What did the one represent? (always will happen) Every time the sticks are tossed, there will be an outcome. The one represents all the possible outcomes.

5. Depending upon time and ability, have the students create a 0-to-1 number line to record the probability for the outcomes in Game Sticks. It will look like this:

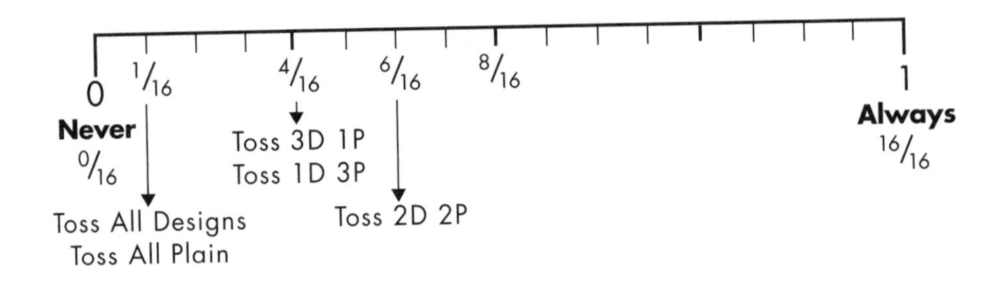

6. Connect the theoretical probabilities to the game. For example, a toss of ALL DESIGN has a probability of occurring $^1/_{16}$ time. Is $^1/_{16}$ closer to zero or one? How close is it to one? What does that tell about the likelihood of that toss occurring?

7. Next link the theoretical probability for TWO DESIGN and TWO PLAIN to the game. Where is it situated on the number line? Compare it to the probability of an ALL PLAIN toss and a THREE PLAIN and ONE DESIGN toss.

8. Continue with the outcome for THREE DESIGN and ONE PLAIN. That toss has a theoretical probability of occurring 4/16 times and receives zero points in the game. That *particular* toss is less likely than a TWO DESIGN and TWO PLAIN toss.

9. Note that the THREE PLAIN and ONE DESIGN outcome also occurs $^4/_{16}$ times and receives zero points. Point out that while separate outcomes, in the game they both net zero points. Connecting this to the game, there is a $^4/_{16}$ + $^4/_{16}$ or $^8/_{16}$ chance of getting either of these outcomes and receiving zero points. In terms of the game, this is the most likely outcome—an outcome of 3 and 1—without attention to the sides that are facing up.

10. End the session with a writing prompt.

■ **Writing Prompt** (For all students)

Use one of the prompts below or write your own for students to respond to. It can be assigned at the end of the session, during journal writing time, or as a homework assignment.

Game Sticks with Three Sticks (easier)
If you only had three sticks to play Game Sticks, what are all the possible ways the sticks could land?
Be sure to clearly list all the possibilities of design and plain sides.
Explain how you know that you have all the possibilities.
How would you assign points for this game?

See pages 118–122 for more information on outcomes for Game Sticks.

Game Sticks with Five Sticks (more challenging)
If you had five sticks to play Game Sticks, what are all the possible ways the sticks could land?
Be sure to clearly list all the possibilities of design and plain sides.
Explain how you know that you have all the possibilities.
How would you assign points for this game?

■ Going Further

Two-Penny or Two-Stick Toss
Students play another game with a two-sided probability tool—either two sticks or two pennies—and assess its fairness.

Players toss pennies or sticks and accumulate points based on the outcomes.
The first player who gets 10 points wins the round.
Players are assigned an outcome to receive points.
They take turns tossing the pennies or sticks.
Points are won after each toss by one of the players, **regardless** of who tossed the pennies or sticks.
The point system is as follows:

Player X gets a point every time 2 HEADS (or 2 DESIGN) are tossed.

Player Y gets a point every time 2 TAILS (or 2 PLAIN) are tossed.

Player Z gets a point every time 1 HEAD (or 1 DESIGN) and 1 TAIL (or 1 PLAIN) are tossed.

Have the students "count off" teams using the letters, X, Y, and Z.
Teams of three (X, Y, Z) play several rounds of the game.
The question to explore: Is this game fair for all players? Explain why or why not.

Design a Game
Have students select a probability tool and design a game. Then provide time for classmates to try out the games and determine if they are fair or not.

TRADITIONAL DESIGNS from CALIFORNIA

Native American Game Sticks

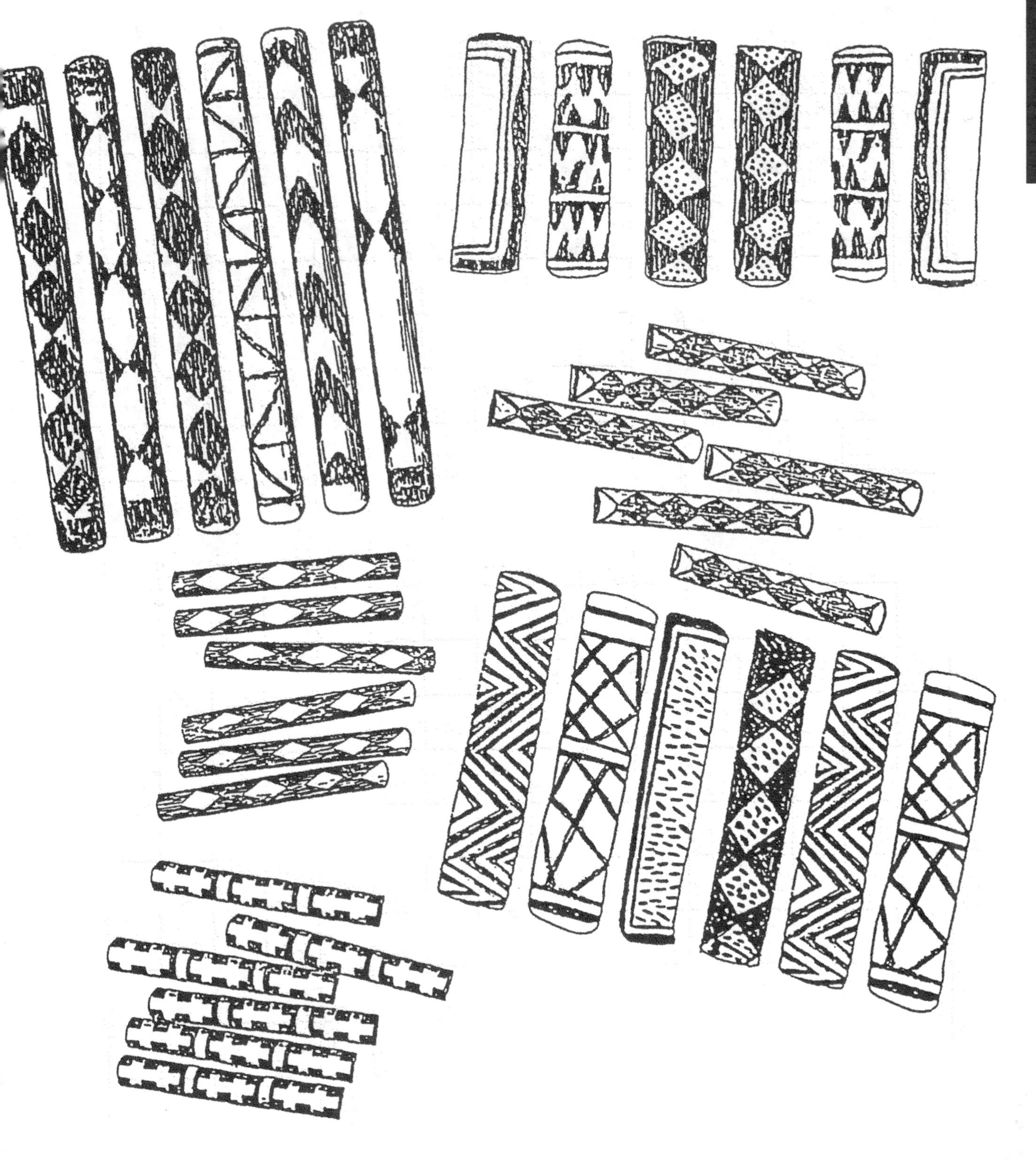

Teachers learn about In All Probability *at a workshop presented by the GEMS Center in Japan.*

Background for the Teacher

Note: This section is a resource for you to refresh your knowledge of the mathematics in the activities in the unit and to help you lead mathematical discussions with your class. It is NOT meant to be read out loud to your class or duplicated for students. As in all GEMS activities, the activities—including the discourse that grows from them—are primary, and are designed to help students discover the concepts for themselves through your facilitation. This content background can also assist you in helping students explore probability in greater depth—especially in response to students' questions and/or conjectures.

The probable is what usually happens. — Aristotle

Aristotle's quotation, written hundreds of years before any historically recorded formal mathematics of probability, is one way to think about probability without using numbers. It serves as a good intuitive definition, and one that you may want to share with your students.

Who Uses Statistics and Probability?

Children first meet probability in the games they play, particularly those using dice or spinners. They hear predictions on weather reports and see statistics on sports cards and cereal boxes and in newspapers.

As adults, we encounter probability in games we play, weather predictions, and economic forecasts. We use statistics to help us decide what new car to purchase, which foods are lower in fat than others, and whether our candidate is likely to win.

Many careers involve statistics and probability. Some examples include accounting, stock brokerage, insurance, real estate, public opinion polling, urban planning, advertising, psychology, elementary school teaching, public health, meteorology (the study and prediction of weather and climate), epidemiology (the study of epidemics, which involves both gathering data and predicting the future course of outbreaks), and seismology (the study of earthquakes, which includes using historical as well as geological data to predict future quakes).

Students' (and our own!) understanding of probability evolves and develops over time and in a variety of contexts. Depending on the developmental level of your students, some may still believe that the results of the flip of a coin or the toss of a die are governed by luck. Others will be able to understand the theoretical probability of an event. Regardless of your students' skills and abilities, *In All Probability* provides concrete experiences with data analysis and builds the foundation for understanding theoretical probability.

Many teachers have found that they learn a lot about probability while teaching their students. Don't worry that you don't know it all yet—plunge in and learn along with your students!

What Are Probability and Data Analysis?

Probability is the area of mathematics that measures the chance or likelihood that something will (or will not) happen. Knowing the probability of an event can help predict an outcome or future event.

Data Analysis is the collection, organization, representation, analysis, and interpretation of numerical facts and data. As students learn in this unit, the collection and analysis of data can provide much information about the probability of something happening.

Vocabulary for Data Analysis and Probability

Theoretical Probability is the probability of an event occurring, based on a mathematical computation. It is often expressed as a number from 0 to 1 inclusive. The probability of "an event" is expressed as a fraction as follows:

$$\text{Probability (event)} = \frac{\text{the number of ways a specific event can occur}}{\text{the total number of possible outcomes for all such events}}$$

The theoretical probability is often referred to as *the odds.*

For example, using the probability of rolling a 4 with a standard die is

$$\text{Probability (roll a 4)} = \frac{1 \text{ (way to get a 4)}}{6 \text{ (total possible outcomes when rolling 1 die)}}$$

"Pr" is used here as an abbreviation for probability.

The smallest possible probability is 0. The probability of rolling a 7 using a standard die is 0. Pr (roll a 7 on standard die) = $0/6 = 0$

The greatest possible probability is 1. The probability of rolling a whole number greater than 0 and less than 7 and on a standard die is 1.
Pr (roll # $<$ 7 on standard die) = $\frac{6}{6} = 1$

The probability of an event happening added to the probability that it will not happen is equal to 1.
For example, the probability of rolling an even number (2, 4, 6) with a single die ($\frac{3}{6}$) added to the probability of rolling a not-even or odd number (1, 3, 5) with a single die ($\frac{3}{6}$) is $\frac{6}{6} = 1$.

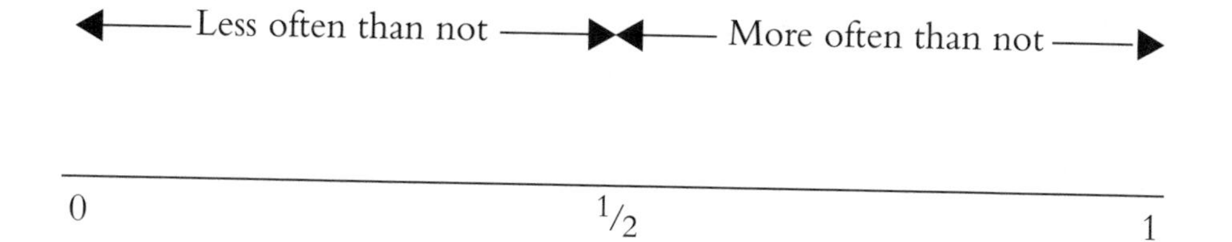

Impossible Unlikely Equally Likely Likely Certain

Sample Space is all of the possible outcomes of an event.
Using the roll of a standard die, the sample space is 1, 2, 3, 4, 5, and 6.
Using the roll of two standard dice, the sample space is 2, 3, 4, 5, 6, 7, 8, 9, 10, 11, and 12.

Equally Likely Events are events that occur with *equal probability.*
Rolling a standard die, there are six possible outcomes—1, 2, 3, 4, 5, and 6. The probability of any one of these six outcomes is $\frac{1}{6}$; therefore, it is equally likely that any one of the numbers will be rolled. No number is more likely or less likely to result than any other number.

Experimental Probability is the probability based on the results of experiments, rather than on what would happen theoretically.

A **Trial** is a single sampling of an event in a probability experiment. For example, one roll of a standard die is a trial.

Sample Size is the number of trials that are conducted in a probability experiment before making a prediction based upon all the trials.

The **Law of Large Numbers** states that the more trials you conduct or data you collect in an experiment, the closer the probability will get to the theoretical probability. A small number of trials may or may not give results very close to the theoretical probability.

Independent Events are experiments or events that do not affect each other. For example, if you toss a penny and you get heads, the probability of getting heads or tails on the next toss is still the same, because each toss is an independent event. When students toss the penny 10 times, each event is independent.

Prediction

Probability often involves prediction. If you toss a penny, will you get a head or a tail? The word *predict* combines two Latin roots. *"Pre"* means *before* and *"dico"* means *say*. So *predict* literally means "say before."

A prediction is a type of guess; there is no right or wrong answer. Depending upon how much information is given beforehand, a prediction may be a "wild guess" (predict how many fish there are in the sea) or an educated guess (predict which horse will win the "Double Dice Derby"). The more information you have to base your prediction on, the closer you can come to the actual number or the probable result.

Both experience and learning prediction strategies will help your students make more realistic predictions. Like any skill, prediction improves with practice.

Data Representation

Data collection, organization, representation, and interpretation are key elements in this unit. Students collect data from objects and through experiments with pennies, spinners, dice, and game sticks (two-sided data generators). They use tally marks, tables, and charts to keep track of data and then display class results graphically.

A graph is one of the most powerful tools in statistics. Graphs are used in mathematics, science, social studies, and other fields as visual means to represent numerical data or thinking and to communicate information. The ability to construct and interpret graphs is vital to mathematical literacy.

Notes on Graphing

Often students are given graphs to interpret without context. They have not generated the data, nor thought about how to present it. One of the goals of this unit is to help students develop the ability to create a graph with an appropriate scale and label it to represent the data they collected. Since this will be new to many students, allow plenty of time to make and discuss various graphs. This time is well spent, as students gain valuable skills in constructing and interpreting graphs

There is a black line master of grid paper (p. 34) for students to use to make graphs as needed. In each activity, graphs are a key component in analyzing the data from the experiments. Lead a class discussion about different ways of graphing and representing data, using graphs created by the students or clipped from newspapers and magazines. A rich discussion of various graphing approaches can result. For example, students may discover that certain methods of recording data make comparison between results simpler, easier to read, etc. As you are in the best position to decide how to effectively instruct and challenge your students, modify the graphing as is appropriate.

What is a Graph?

A graph is a display of data to compare information or to show a relationship between sets of data. A graph has a **pair of axes** (horizontal and vertical) that need to be labeled with names and/or numbers to provide meaning for the data. A graph also needs **a title** and **date.**

The **scale** of the graph needs to be determined based on the range of the data and the size of the graph paper. The difference between the numbers from one grid line to another is called the **interval.** It is important to check the scale of a graph, because the scale can skew your interpretation of the data.

Interpreting Graphs

When reading graphs, it is important to carefully examine how the data is represented. Once that is understood, the factual information displayed on the graph can be identified. These **facts** are **true statements** about the data. Some examples of true statements students might generate about the Penny Flip graph are:

We got 192 tails and 178 heads in our experiment.
There were more tail outcomes than head outcomes.
There was a total of 370 penny flips in this experiment.
There were 14 more tail outcomes than head outcomes.
The number of head and tail outcomes in our experiment was relatively close.

An ***inference*** is a statement based on **interpretation** of the data on the graph. The interpretation may not accurately reflect what the data is showing. In advertising, data is often displayed in a way that promotes products or services. The graph comparing motorcycle brands (p. 32) is an example of that. Some examples of inferences students might make from the Penny Flip graph are:

Tails is the lucky side of the penny.
You are more likely to get a tail when you toss a penny.
When you toss a penny, you seem equally likely to get a head or a tail.

As illustrated in these examples, some inferences reflect an understanding of the experiment and the outcome while others are inaccurate conclusions.

Types of Graphs

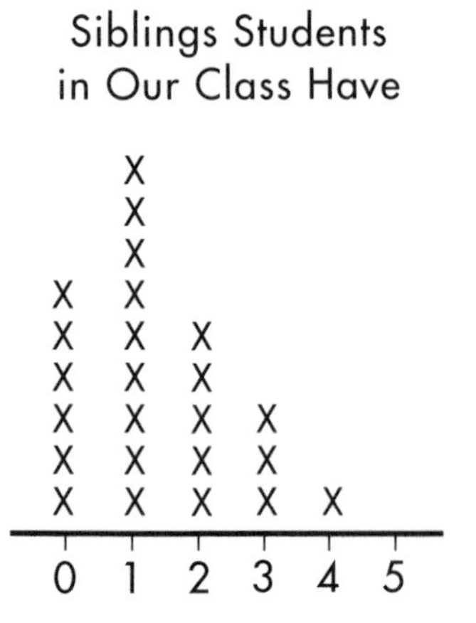

A **line plot** uses a number line to display data. Each piece of data, represented by an "x," is marked above the corresponding number on the line. It groups or clusters the data.

A **bar graph** is used to compare data. The bars can be situated vertically or horizontally. The height or length of the bar represents a number. For example, the graph of the number of heads and tails tossed is a bar graph.

A **histogram** is similar to a bar graph except that each bar represents an interval, and there are no spaces between the bars.

A **double bar graph** has pairs of bars that compare sets of data. For example, the outcomes of the tosses of the game sticks from two different games can be compared on this kind of graph.

A **pictograph** uses pictures or symbols to represent and compare data. For example, one penny can represent 5 head outcomes in a penny toss activity.

A **circle graph** represents all the data collected within the area of the circle by dividing the circle into fractional parts. For example, after rolling a die 100 times, the outcomes can be represented for each number as a fractional part of the whole circle.

A **line graph** is used to show how two pieces of information are related and how they vary depending on one another. When you look at the line in a line graph, you can tell whether something has increased, decreased or stayed the same. *Note:* A line graph is *not* the graph of a line.

Survey Graphs

If your students are unfamiliar with graphing, you can give them practice by having them take surveys and create bar graphs with the results. Have students ask members of the class (or the school or some other group) a question on a particular topic, collect the responses, and then represent the data. For example, students might conduct surveys to find out the current favorite music group, the most common method students use to get to school, or the languages other than English spoken by the students in your class and their families. Your students will surely come up with their own ideas to generate questions and create survey graphs.

The website http://www.edhelper.com/graphs.htm *has activities and data sheets to support students in creating graphs.*

A Tip on Data Collection: Suggest that students limit the choices they give when they conduct their surveys. For example, they may give four music groups, and ask each student to choose his or her favorite among those. Otherwise, if the choice of group is left open-ended, they may end up with thirty kids choosing thirty different musicians. That graph would not lend itself to determining scale or making any significant interpretations.

You may want to do one survey with the class to demonstrate how to collect the data, create a bar graph, enter the data, and then make some interpretation of it. For example, you could ask students to select their favorite fruit from among four choices (banana, apple, tangerine, and watermelon). Demonstrate how to tally and count the results to determine the scale. Then make a graph with labels, a title, and the data. Post the model graph for reference as students make their own.

Into the Classroom

Discourse: The Teacher and Student Roles

Discourse is the way knowledge is constructed and exchanged in the classroom. Both the teacher and the students play important roles in shaping the discourse. Above all, mathematical discourse is focused on making sense of mathematical ideas and using those ideas in logical ways to set up and solve problems and make evidence-based explanations.

Teachers serve as skillful facilitators who encourage and provide context and structure for the communication and learning of mathematics by posing questions, listening carefully to students' ideas, guiding discussions, allowing students to grapple with problems, providing content, and monitoring participation of all students. Skillful facilitation involves daily professional growth, with continual development of pedagogical content knowledge, recognizing "teachable moments," and knowing which interventions (if any) to use and the most strategic points at which to use them.

Students are active participants in their learning, working both collaboratively and independently. They should have access to a variety of tools to solve problems and present solutions. Students also make conjectures and try to convince one another of the validity of representations, solutions, and answers. They rely on mathematical evidence and arguments to determine validity.

The classroom environment is a major influence on what students learn in mathematics. For discourse to flourish, the environment needs to support open dialogue and engage all students in mathematical thinking and learning. Students are an audience for each other. To enhance discourse, the teacher encourages the use of concrete materials as models; pictures, diagrams, tables, and graphs; invented and conventional terms and symbols; written and oral hypotheses, explanations and arguments; and technology tools.

Throughout this guide, discourse is used to introduce and develop mathematical ideas. Students actively construct understanding of concepts—from the concrete to making connections on more abstract levels—through activities that build on prior knowledge. Multiple ways of gathering, representing, and analyzing data provide new tools and perspectives to apply when encountering similar problems.

Additional information on mathematical discourse with examples of discourse in action can be found in the Professional Standards for Teaching Mathematics *(see "Resources" on page 135).*

Literacy and Math

As the mathematics content is presented and developed in this unit, literacy skills are built upon and advanced. Key mathematics vocabulary and concepts are introduced to enable students to articulate ideas and solutions using appropriate language both orally (speaking and listening) and in writing. Distinction is made between the standard meaning of words and their mathematical use. For example, a table is commonly known as a piece of furniture upon which objects can be set. In math, a table is a tool for organizing data. The language developed in the unit is recorded in the classroom to serve as a reference for all students in their work. Writing can be integral to the unit and provide a window on student thinking and understanding.

Into the Mathematics, Activity by Activity

Activity 1: *Penny Investigations*

Students flip a penny and record the results. In the case of flipping one penny, there are two possible outcomes: the penny will land on heads or it will land on tails. It is so unlikely that the penny would stay in the air or land on its edge that either of these is not considered a possibility. The penny is "fair"—when you flip a penny, it is equally likely that it will land on heads as on tails.

The theoretical probability of tossing a penny and getting a head is expressed as the number $^1/_2$. The same is true for getting a tail. This theoretical probability can also be represented as a decimal, 0.5, or a percentage, 50%. You can also say that there is a $^{50}/_{50}$ chance of getting tails or heads.

What is the probability of getting either heads or tails on your next flip? Since heads and tails are the only two possible outcomes, you will surely get one on the flip. Therefore, the probability can be expressed as a 1—it will always happen. Similarly, it can be said you have a 100% chance of getting a head or a tail on a penny flip.

What is the chance of getting a Jefferson head when you flip a penny?
This is not a possibility; you will never get a Jefferson head when you toss a penny, as pennies are currently minted with Lincoln on the head side. Therefore, the probability can be expressed as a 0—it will never happen. Similarly, it can be said that you have a 0% chance of getting a quarter when you toss a penny.

What is the chance of getting a head after tossing a head? The chance of getting a tail after tossing a head?
Each toss of a penny is an independent event. Therefore, each toss does not affect the next toss. On each toss, there is a theoretical probability of $^1/_2$ for getting a head or a tail. It is equally likely you will get a head or tail.

Since there is an equally likely chance of getting a head or tail, why are results from the 10 tosses experiment not always 5 heads and 5 tails?
When your students toss their pennies 10 times, it would seem logical to get 5 heads and 5 tails. However, such a small sample often produces results that differ quite a bit from the theoretical probability of tossing a penny. That is why students flip 10 more times each, and then the class compiles all their data. As the sample gets larger the results are likely to be a close approximation of the theoretical probability.

Students may start to see this phenomenon on their own as they participate in the activities of this unit—their own results as well as their results combined with their partner's (a small number of trials) may not be close to the theoretical probability, while the class results (a larger number of trials) are likely to be quite close. Guide students into noticing that the class data more closely reflects the theoretical probability.

Penny Information

Pennies have been minted in the United States since 1793. Half pennies were minted from 1793 to 1857. If you are interested in learning more about old coins, visit the Web site http://www.coinfacts.com for a wealth of information on pennies, as well as other U.S. coins. For a modified history of pennies, visit http://www.pennies.org/history/one.html

For a downloadable coloring sheet of the current penny, visit:
http://www.enchantedlearning.com/math/money/coloring/penny.shtml
By clicking on the word *Penny* on that page, you will find information and additional activities on pennies for your students.

The following is a brief history of the evolving penny from the late 19th century to date.

- Flying Eagle Cent 1856-1858: The Flying Eagle Cents were struck without official authorization, and therefore all 1856 Flying Eagle Cents are considered illegally struck. This coin represented a transition from Large Cents. It cost more money to manufacture than its value.

- Indian Head Cent 1859-1909: The Indian Cent was first introduced in 1859 and depicted an Indian princess on the obverse (head side). A popular story about its design claims a visiting Indian chief lent the designer's daughter his headdress so she could pose as the Indian princess.

- Lincoln Head with Wheat Ear Reverse 1906-1958: The minting of this penny was the first time a portrait coin appeared in our regular series. A strong feeling had prevailed against the use of portraits on the coins of the country, but public sentiment stemming from the 100th anniversary celebration of Abraham Lincoln's birth proved stronger than the long-standing prejudice. The metal used to mint the coin changed over the years as follows:

  1906-1942 copper
  1943 only year with zinc-coated steel due to WWII
  1944-45 bronze
  1946-1958 copper

- Lincoln Head with Lincoln Memorial 1959-date

  1959-1982 copper
  1982-date copper-plated zinc (the rising price of copper led to a composition change—to an alloy of 97.5% zinc and 2.5% copper)

Activity 2: *Track Meet*

This is the first activity in which the students experience unequal chances. On Spinners A and B there are three possible outcomes, red, yellow and blue—however, the colors are not equally likely to occur on each spinner.

Spinner B is divided into three sections of equal area; this spinner is designed with equally likely outcomes. There is a one-in-three chance of getting yellow, red, and blue. The theoretical probability is represented by $^{1}/_{3}$. The chance of getting yellow is 0.33 or about 33%.

On Spinner A, the area of the circle is divided such that red and yellow each have two of eight equal parts, and blue has four of the eight equal parts. As a result, it is not equally likely to get each color on a spin. Blue is a more likely outcome than red or yellow. In this case, the theoretical probability for spinning a red is expressed as the number $^{2}/_{8}$ (or $^{1}/_{4}$). It is less likely to get a red than a blue or a yellow combined on a spin. Similarly, the same holds true for the color yellow. For blue, the theoretical probability is expressed as the number $^{4}/_{8}$ (or $^{1}/_{2}$). The chance of getting a blue is equally likely as getting a red or a yellow on any spin.

Spinner C is divided into four seemingly equal areas. However, two of the areas—red and green—are equal and slightly larger than the other two areas—blue and yellow— that are also equal in area. The results from the race will demonstrate that even though a spinner may appear fair to the naked eye, it needs more thorough evaluation to determine its fairness as a probability tool.

The outcome of the races should show that each of the two larger areas had more wins than the smaller areas, though the results may be close! To determine the actual fractional part of the area, the circle has been divided using degrees with red and green, each covering $^{95}/_{360}$ of the circle area (or $^{190}/_{360}$ for both) and blue and yellow each covering $^{85}/_{360}$ of the circle's area (or $^{170}/_{360}$ for both). In percentages, red and green together will win a little more than 52% of the time compared to a little less than 48% of the time for blue and yellow together. Taken separately red and green will each win about 26% of the time; blue and yellow will each win about 24% of the time.

Activity 3: *Horse Racing*

Students investigate the probability for each of the numbers, one through six, when rolling a standard die. As long as the die is not weighted unfairly, each of the numbers has an equally likely chance of landing up on any one toss. If a die is tossed only six times, it is unlikely that each of the six numbers will be rolled one time. *Each roll is an independent event;* therefore each number has a *1 chance in 6* possible chances of landing up on each toss. However, as with the penny toss, the greater the number of trials, the more likely it is that each number will be rolled about the same amount of times as every other number.

Students who play the Horse Race game roll the dice many times in a single game to determine a winner of the race. Often they see the horses neck-and-neck as they move toward the finish line. Since only the winners are recorded, the more games that are recorded, the more it becomes evident that each horse wins a fairly equal number of times. It is *equally likely* for each horse to win a race.

Roll ALL Six presents another good opportunity to help students see the Law of Large Numbers in action. Results from the experiment by each student, and then from pairs of students, are likely to vary greatly from the theoretical probability (that each number will occur about one-sixth of the time). When the class results are combined, the number of times each number was rolled will be about the same. It is *equally likely* for each number to be rolled.

Double Dice Derby is a race among 12 horses in which the sum of two dice "moves" the horse forward, one step at a time. The winning horse of each race is recorded on a large class graph. In contrast to the equally likely chances of getting the numbers one through six with one die, with two dice the *chances of getting each possible sum* (two through twelve) are *not equally likely.* By creating a graph of the race winners, students see that it is not a "fair" race. Some horses win many times, other horses win only a few times, and Horse #1 never moves at all!

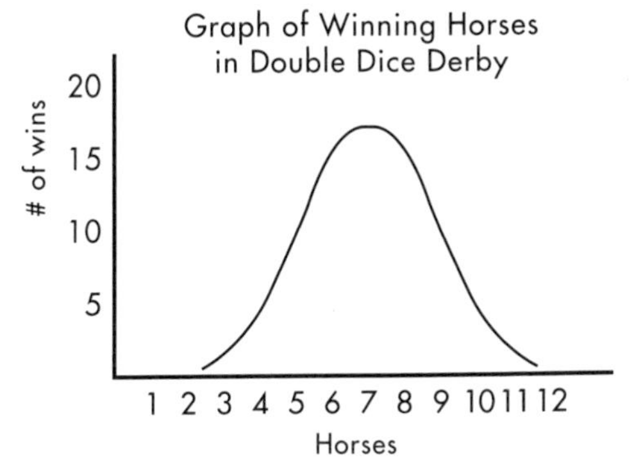

The class graph of winning horses frequently has a *bell shaped—or normal—curve.* The data clusters around the center of the graph toward the normal tendency to get a 7. Similarly, when students play the game, once a horse has won the race, the ending position for all the horses on the game board also often resembles a bell curve.

The "Keeping Track" chart makes transparent the reason why horses six, seven, and eight win so frequently. Each of the 36 possible outcomes for the sum of two dice is represented on the chart. Though there are only 11 possible sums, 2 through 12, some sums appear more frequently than others. The 11 sums are *combinations* using two dice. However, some of those sums can be generated in more than one way! For example, the number seven occurred most frequently on the chart. Seven can be generated as follows: 1+6; 6+1; 2+5; 5+2; 3+4; and 4+3. Those distinct ways—in which order matters—to have a sum of seven are called *permutations.* When playing the game, attention is not paid to order or specific combination—only to the combination's sum.

The following two definitions further explain how combinatorics, a branch of mathematics, plays a role in this activity:

A ***combination*** *is a selection of a set of objects irrespective of the way they are chosen.* In the case of this game, the selection of two numbers is from the 12 possible numbers that can be generated using two standard dice (1, 2, 3, 4, 5, 6, 1, 2, 3, 4, 5, 6). Since it is the sum of the combination of two numbers that is used, the order of the two numbers is not important.

A ***permutation*** *is an arrangement of a set of objects in a definite order.* All of the possible two-number combinations that equal a particular sum are the permutations. For example for a sum of four, there are three permutations: 1+3; 3+1; and 2+2 (of the two combinations "2 and 2"and "1 and 3").

Here is a listing of the permutations for the sums (combinations) generated by two dice:

2	3	4	5	6	7	8	9	10	11	12
1+1	1+2	1+3	1+4	1+5	1+6	2+6	3+6	4+6	5+6	6+6
	2+1	2+2	2+3	2+4	2+5	3+5	4+5	5+5	6+5	
		3+1	3+2	3+3	3+4	4+4	5+4	6+4		
			4+1	4+2	4+3	5+3	6+3			
				5+1	5+2	6+2				
					6+1					

After the students complete the "Keeping Track" chart, have them find the patterns on the chart. Connect the outcome on the chart to the lists they generate. Check that there are exactly 36 possible outcomes when you roll two dice. This is a finite countable number. How frequently a sum occurs determines how likely it is to be rolled. The two representations of the outcomes—the chart and the sums as listed above—can be compared to determine which is easier to read and interpret.

Through this investigation, students should be able to see that it is much more likely for Horses #6, #7 and #8 to win. There are six ways to get a sum of seven, making it most likely for Horse #7 to win the race. However, there are five ways to get a sum of six, as well as five ways to get a sum of eight. It is likely that Horses #6 and #8 will win some of the races.

What if Horse #7 does not have the most wins on the class chart?
There are times that either Horse #6 or #8 will win more times than Horse #7. That does not mean something is wrong with the dice or that the students are cheating! Probability helps us to predict what is most likely to happen, not what will definitely happen.

What if there are wins reported for horses that are not likely to win?
Again, this is possible. Students may have played fairly. Though the chances of Horses #2, #3, #11, and #12 are slim for winning the race, it is possible, though not probable.

Is there a way to make the race "fair" for all horses?
By changing the way the game is played, it is possible to create a game where all players, but not all horses, have an equally likely chance of winning. For example, if three players were racing the horses, they would divide the horses among themselves such that the probability for each person's horses would equal $^{12}/_{36}$ and each player would have a horse that is likely to win. For example: Player A has Horses # 2, 3, 7, 11, 12; Player B has Horses #4, 5, 6, and Player C has Horses #8, 9, 10.

Theoretical Probability for Double Dice Derby

The theoretical probability for each sum on a given roll of two dice is as follows:

Pr (2) = 1 (way that the sum will be equal to 2 **or** 12)
Pr (2 **or** 12) 36 (total possible outcomes when rolling two dice and adding the numbers)

Pr(3) = 2 (ways that the sum will be equal to 3 **or** 11)
Pr (3 **or** 11) 36 (total possible outcomes when rolling two dice and adding the numbers)

Pr(4) = 3 (ways that the sum will be equal to 4 **or** 10)
Pr (4 **or** 10) 36 (total possible outcomes when rolling two dice and adding the numbers)

Pr(5) = 4 (ways that the sum will be equal to 5 **or** 9)
Pr (5 **or** 9) 36 (total possible outcomes when rolling two dice and adding the numbers)

Pr(6) = $\frac{5}{36}$ (ways that the sum will be equal to 6 **or** 8)
Pr (6 **or** 8) (total possible outcomes when rolling two dice and adding the numbers)

Pr(7) = $\frac{6}{36}$ (ways that the sum will be equal to 7)
(total possible outcomes when rolling two dice and adding the numbers)

Activity 4: *Game Sticks*

As in the Double Dice Derby, combinations and permutations play a significant role in this game. To win points, the combination of the sticks, not the order in which they land, is key. The number of counters taken depends upon the combination of design sides and plain sides. To understand why the game unfolds the way it does, let's take a closer look at permutations.

One Stick
To gain insight into the probability of the Game Sticks, it helps to make the game simpler. Begin with just one stick (similar to a penny toss). When one stick is tossed, there are only two possible outcomes—it will land with the design side up or the plain side up. (With the penny, it was either heads or tails.)

Two Sticks
Moving on to two sticks (similar to the Two Penny Toss Game, page 100), there are three combinations that can result from a toss: two design sides; two plain sides; and one design side and one plain side. However, there are four ways to get those three combinations. The design side and plain side combination has an "invisible" outcome if you do not look at the order in which the sticks land! The following graphics represent the *four* ***permutations*** *of tossing two sticks:*

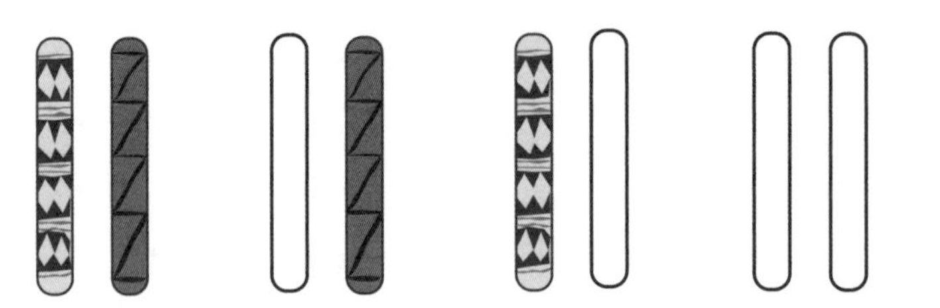

The probability of getting two design sides is one out of four ($^1/_4$) as is the probability for getting two plain sides ($^1/_4$). However, the probability for getting a plain side and a design side is two out of four ($^2/_4$).

Three Sticks

If you use three sticks, there are four combinations that can be tossed: three design sides; two design sides and one plain side; three plain sides; and two plain sides and one design side. Though there are only these four, there are eight distinct ways (permutations) that the sticks can fall to create those outcomes. They are as follows:

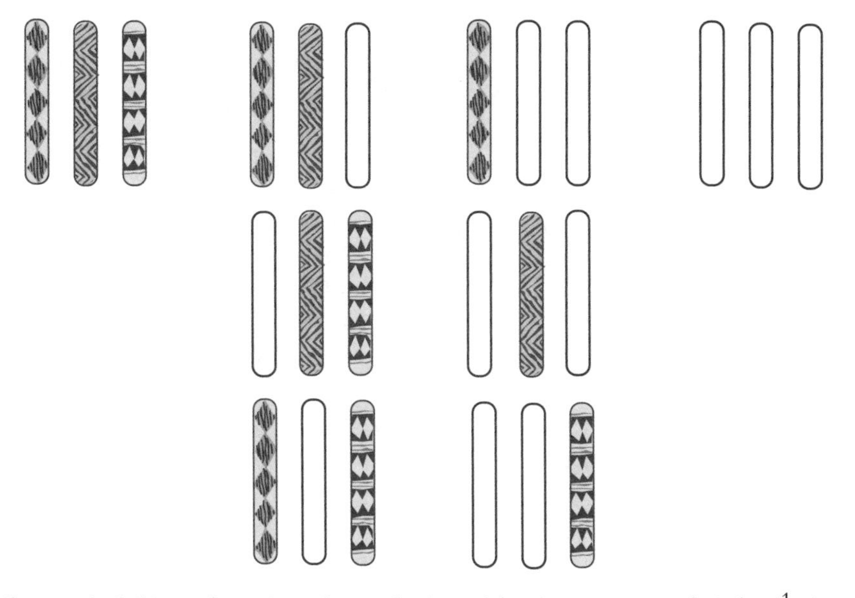

The probability of getting three design sides is one out of eight ($^1/_8$) as is the probability for getting three plain sides ($^1/_8$). The probability for getting two design sides and one plain sides is three out of eight ($^3/_8$) as it is for getting three plain sides and one design side ($^3/_8$).

Four Sticks

This is the version of Game Sticks that your students play. Using four sticks, there are five combinations in which the sticks can fall—4 design sides; 3 design/1 plain; 2 design/2 plain; 1 design/3 plain; and 4 plain sides. However, there are sixteen permutations of these combinations!

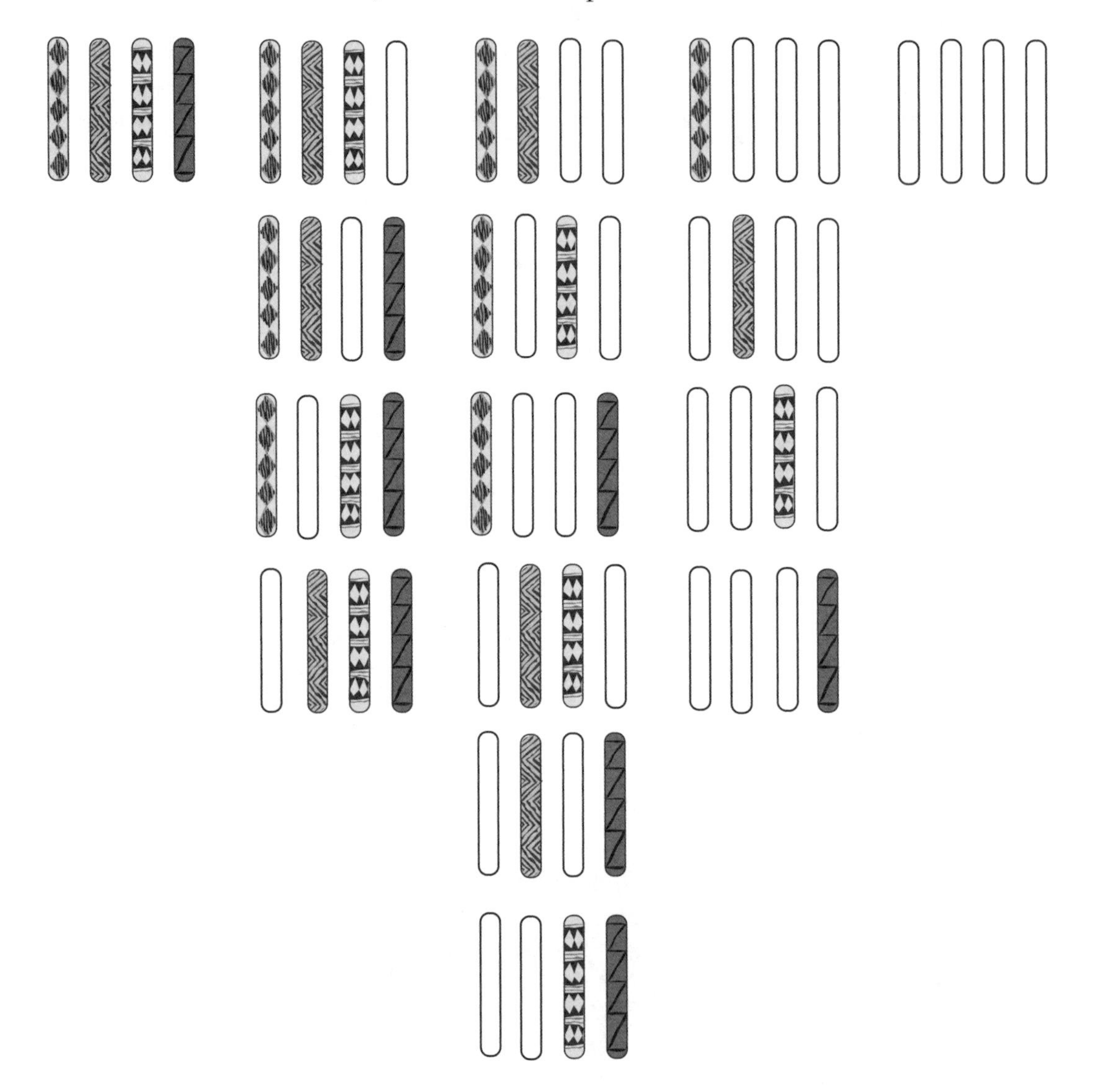

Notice that there is only a 1-out-of-16 chance of getting all design sides up, but a 6-out-of-16 chance of getting two design and two plain. The theoretical probabilities for the outcomes are as follows: $^{1}/_{16}$ for getting all design OR all plain sides; $^{4}/_{16}$ for getting 3 plain and 1 design OR 3 design and 1 plain; and $^{6}/_{16}$ for getting two (or half) design and two plain.

Native American Game Sticks

Many Native American cultures included games of chance in their diverse recreational activities. In California, during the winter rains, it was common to play dice games, and many tribes throughout the Americas also played versions of dice games. Both men and women played games of chance, often separately. As men played dice games, songs and chants were often sung to help bring luck to the players, or, on the other extreme, the game was played in silence. When women played dice games, there could be conversation and joking rather than singing.

The game pieces were made from a variety of local materials, including split canes, wood, bones, beaver or woodchuck teeth, walnut shells, peach and plum stones, and sea shells. The dice, or staves, would have two faces—one side marked with patterns or designs and the other side plain.

There were two main methods used to keep track of the score--one in which the score was kept with sticks or counters, and the other in which a counting board was used. The counters were usually in sets of 10 and were passed, hand-to-hand, until one person (or opposing side) had all the counters. There is evidence that some tribes used 12 counters. The counting board consisted of a square or circle made with stones on the ground, or was an inscribed board made of cloth or hide. The board was used to move "men" and indicate the position of each player.

Game Sticks in Activity 4 is modeled after the dice games traditionally played by Native American women who lived in what is now California. The women would generally use six staves to play the game. Adding two more sticks increases the number of combinations to seven, but the number of permutations grows to 64! If your students are ready to see how the game changes with six sticks, have them combine their sticks to have one set of six sticks to play with.

For your information, on the next page are the combinations and permutations when six sticks are used.

Combinations and Permutations for Six

Native American Game Sticks

■ A = Design Side
□ B = Non-Design Side

SIX STICKS

Combinations	Permutations
■■■■■■	AAAAAA
■■■■■□	AAAAAB
	AAAABA
	AAABAA
	AABAAA
	ABAAAA
	BAAAAA
■■■■□□	AAAABB
	AAABAB
	AABAAB
	ABAAAB
	BAAAAB
	AAABBA
	AABABA
	ABAABA
	BAAABA
	AABBAA
	ABABAA
	BAABAA
	ABBAAA
	BABAAA
	BBAAAA

SIX STICKS

Combinations	Permutations
■■■□□□	AAABBB
	AABABB
	AABBAB
	AABBBA
	ABAABB
	ABABAB
	ABABBA
	ABBAAB
	ABBABA
	ABBBAA
	BAAABB
	BAABAB
	BAABBA
	BABAAB
	BABABA
	BABBAA
	BBAAAB
	BBAABA
	BBABAA
	BBBAAA

SIX STICKS

Combinations	Permutations
■■□□□□	AABBBB
	ABABBBB
	ABBABB
	ABBBAB
	ABBBBA
	BAABBB
	BABABB
	BABBAB
	BABBBA
	BBAABB
	BBABAB
	BBABBA
	BBBAAB
	BBBABA
	BBBBAA
■□□□□□	ABBBBB
	BABBBB
	BBABBB
	BBBABB
	BBBBAB
	BBBBBA
□□□□□□	BBBBBB

ACTIVITY 1: PENNY INVESTIGATIONS

Session 1: Introducing Data

■ Getting Ready

1. Students should have math journal or pocket folder.
2. Gather chart paper, markers, pennies 1970s to present.
3. Prepare graph for organizing pennies by year.
4. Familiarize yourself with Background for Teachers.
5. Send letter about unit home.

■ Data, Data, Everywhere!

1. Write the word data on board and ask about meaning and use. Record ideas.
2. Give a penny to each student to identify, discuss value. Circulate, listen, have class share observations.
3. Ask about, record, and graph oldest pennies.
4. Have students talk to partners about graph. Distinguish factual data from inferences.
5. Assign homework.

Session 2: Penny Flip

■ Getting Ready

1. Record outcomes when penny flipped 10 times on cards, as in guide.
2. Have pennies from Session 1 available.
3. Fill a small jar with up to 59 pennies.
4. Gather two pads of post-its in contrasting colors.

■ Pennies from Home

1. Students share pennies from home with partner. Who has oldest? Ask dates of oldest ones, record, and determine which is oldest.
2. If no one found out as homework, say pennies first minted 1787 in Philadelphia, pure copper, designed by Benjamin Franklin.

■ Estimation Jar

1. Hold up small jar of coins. Ask students to estimate how many pennies, talk with partner, share estimates.
2. Pour pennies out of jar; students revise estimates.
3. Count out 10. Look at those remaining. Are there 100 pennies? Count out another 10. Continue until 40 counted. Have students make final estimation.
4. After penny amount known, have students reflect on how estimates changed. What other coins would equal the amount of money in jar? Encourage multiple answers.

■ Penny Flip

1. Have they flipped a coin? Why do people flip coins? What outcomes possible? Is it a fair way to decide? Why or why not?
2. Write PREDICT on board (from Latin, to say before).
3. They will flip a penny 10 times. Ask for possible outcomes and list.
4. Select a student partner and model how to conduct experiment.
5. Give one post-it per student and guide them in setting up recording system.
6. Distribute pennies and remind students to predict and record predictions in journals.
7. Post index cards with outcomes. Circulate and assist as needed.
8. How close were predictions? Were they surprised? Why?
9. They have results from two experiments. Is this enough to predict what is likely to happen when you toss a coin 10 times? For more information, they will combine the data generated by whole class and graph. Ask for observations and pose questions.

■ Penny Flip Revisited

1. Partners conduct experiment again. Remind them to predict outcome before they begin.
2. Circulate. After they finish, ask them to predict what will happen as more results are added.
3. Continue as in guide, with students posting results on class graph, predicting, writing final predictions in journals, and explaining reasoning.

Session 3: Making "Cents" (Sense) of the Penny Data

■ Getting Ready

1. Make three overheads of the grid.
2. Make one paper copy of grid for each student.

■ Calculating the Number of Heads and Tails

1. Students review what they wrote about number of heads and tails flipped by class. They work together to calculate total heads and tails.
2. Guide students in creating a grid to record outcomes.

■ Creating a Class Graph

1. Ask students to discuss a way to graph total number of heads and tails.
2. Project a cm grid. Have students propose a scale; work with them to finalize. Emphasize labeling, key, title.
3. Ask students what factual information the graph provides.
4. Are they equally likely to get a head as a tail? Why or why not?
5. For homework, have students record total number of heads and tails in journal.

Session 4: Graphing Heads and Tails

■ Getting Ready

1. Gather and clean overheads from Session 3.
2. If using graph with incorrect scale, make overhead.
3. Make overhead of motorcycle brands graph. Collect graphs from other sources.

■ How Scaling Affects the Graph of the Data

1. Put class graph from the prior session on overhead.
2. Have students share graphs they made for homework with groups or partner.
3. How many students used same scale as class graph? Did anyone use a different scale? Select a student's graph to discuss. As student explains scale, record it on a grid overhead. Continue discussion of graphing and scale.
4. Reinforce idea that the data collected informs the scale of a graph.

■ Truth in Advertising

1. Project motorcycle graph.
2. Have partners discuss graph, then have class discussion. Discuss the scale. Emphasize that the intent is for viewer to INFER that Brand A is best.
3. Have students discuss why it is important to be able to read graphs.

■ Graph Again!

1. Have students make one more graph, with options as in guide.
2. Have students look for graphs at home and complete homework assignment.

Session 5: Introducing Theoretical Probability

■ Getting Ready

1. Decide if you will read Cloudy With a Chance of Meatballs.
2. Gather chart paper and pens. If you decide to use a tool to measure the point $1/2$ on the 0-to-1 number line, have string available the same length as number line.

■ Always and Never

1. Read *Cloudy With a Chance of Meatballs.* Discuss. Is it really possible to have weather come in the form of food?
2. On erasable board, draw number line from 0 to 1. Explain use of 0 as never and 1 as always.
3. Have students discuss examples of events that could never happen, that always happen, that sometimes happen.

■ Connecting to Probability

1. Have students think back to penny flip. Did it ALWAYS come up heads? NEVER heads? ALWAYS tails? NEVER tails?
2. Reexamine outcomes of penny toss. It is EQUALLY LIKELY to get a head or a tail. There is a one-out-of-two chance for a head or a tail. It can be expressed as the fraction one-half (1/2).
3. Draw new number line from 0 to 1 on chart paper. Ask where 1/2 would be on the number line. Have a student suggest location. Check if others agree. Label point that represents 1/2.
4. Is penny toss a fair way to decide something? Why or why not? When class agrees it is fair, below 1/2 write "equally likely."
5. Are there other examples of equally likely events? Have students discuss and add examples.
6. Provide writing prompt. Collect journals to assess understanding and inform your teaching of next activity.

ACTIVITY 2: SPIN TO WIN

Session 1: Ready, Set, Spin!

■ Getting Ready

1. Make cardstock copy of Track Meet board for each student pair, one overhead copy, and two copies on white paper.
2. For each pair, make one Spinner A and one Spinner B, make overheads, gather markers and crayons, make graphs as detailed in guide.
3. Duplicate copy of the homework sheet and make one overhead copy.

■ Track Meet

1. Ask students if they've heard of or seen a track meet. They will be simulating a track meet among three "runners."
2. Show board on overhead. Show starting and finishing lines and runners. Explain game and demonstrate with a student. They will conduct six races, three with each spinner.
3. Circulate and assist as necessary. When students finish, ask questions to spark discussion.

■ Graphing Track Meet Winners

1. Post graphs, "Winners of Track Meet Using Spinner A" and "Winners of Track Meet Using Spinner B."
2. Record all winners for Spinner A and create bar graph, then for Spinner B.
3. Have partners discuss both graphs. Ask questions about the "fairness" for each runner. Come to consensus that Spinner A appears unfair and Spinner B is fair.
4. Put homework on overhead and explain assignment.

Session 2: What's the Real Spin on the Spinners?

■ Getting Ready

1. Create two graphs to record where each runner is situated at the end the third race.
2. Cut the completed Track Meet data sheets into strips by color.
3. Duplicate master copies of Spinner A and Spinner B on regular paper and prepare as in guide.
4. Have crayons available in yellow, red, and blue. Students will also need scissors.
5. Read the Background for Teachers section on Law of Large Numbers and decide what prompt you will use to assess their understanding of the fairness of the spinners.

■ Winners vs. Distance Covered

1. Provide time for students to share their homework in small groups.
2. Focus attention on graphs of Track Meet winners using Spinners A and B. Ask what information graph provides. Have students look at the spinners in relationship to their graphs. Does the spinner help predict the winner? Why or why not?
3. Say you have cut up data sheets with results of third races using Spinner A and B. Distribute results of Spinner A to one partner and of Spinner B to the other.
4. Put up graph "Distance Covered by Runners Using Spinner A." Have students tape results on graph. Start with yellow and proceed as in guide. Ask for observations.
5. Create a graph for "Distance Covered by Runners Using Spinner B."

■ Parts of the Whole

1. Is Spinner A a fair spinner? Why or why not?
2. Distribute copy of spinner. Have them color in each section then cut out the pieces.
3. Compare size and shape of each piece. They are all equal! Write the theoretical probability each color represents. Tell students about Law of Large Numbers.
4. Distribute copy of Spinner B to each student to color in and cut out. Have them make a circle by placing all the yellow, red, and blue pieces together in groups. Discuss.
5. For students in Grades 4 and 5 discuss theoretical probability and placement on number line.
6. End with writing prompt.

Session 3: Spinner Investigation: Fair or Not Fair?

■ Getting Ready

1. Review homework from Session 1.
2. Use Track Meet Rematch master to prepare copies.
3. Use Spinner C master to prepare spinners.
4. Gather markers used in Session 1 and a green marker.
5. Gather crayons used in Session 2 and a green crayon. Have scissors available.
6. Create two graphs to record the winners of races and distances covered.
7. Duplicate "Fair or Unfair?" assessment and make overhead.

■ Track Meet Rematch

1. Tell students you have a new spinner and need their help to determine if it is FAIR or UNFAIR. Show on overhead. Have students discuss its fairness with partner or in groups.
2. Show the board. There are four colors racing. Review the procedure.
3. Circulate as students play and record the winners of each race. When all races are complete, focus on "Winners" graph. From the data, is the spinner fair? Unfair? Why or why not?
4. Post "Distances Covered by Runners in the Rematch" graph. Have students report number of spins for each color then record data on graph. Shade in one box for each spin.
5. Start with yellow and continue color by color until the graph is complete. Does this data provide any new insight? Why is this spinner fair or not?
6. Distribute copy of spinner to each student. Have them cut the spinner into its color pieces and compare size of each. Ask if pieces have equal area. If so, which ones? Connect the fractional parts of the spinners back to the results on the graphs.

■ Assessing Understanding

1. Distribute "Fair or Not Fair" data sheet for students to complete independently.
2. Provide about five minutes for students to work then collect papers.
3. Project overhead of data sheet. Have students talk to partners, then have a class debate about the fairness of each spinner.

ACTIVITY 3: HORSE RACING

Session 1: Roll of a Die

■ Getting Ready

1. For Sessions 1, 2, and 3, you will need at least two dice per pair of students.
2. Read through both experiments, Horse Race and Roll ALL Six, and decide which you will do. Follow preparation steps for the one you choose.

■ Introduction to a Die (for both experiments)

1. Ask students if they've ever used dice. Have them do a "quick write" in their journals of what they know about dice and how dice are used.
2. As students share, record their prior knowledge. Build on this to define a standard 1-through-6 die. Do they think a die is a "fair" tool, like a penny, to determine something?
3. Say they are going to use a die to conduct an experiment.
4. Continue with either Experiment 1 or Experiment 2, as detailed in guide, explaining procedures, recording results, and stimulating discussion.

■ Theoretical Probability of a Die

1. Continue discussion using the theoretical probability of rolling any one number on a standard die. Have students recall that a die has six square faces. There is one number on each face and that number is different on each face.

2. In every case, it is one chance out of a possible six. Because each number has the same chance, it is fair. However, it's not likely they will roll the numbers one through six in six rolls! Each time the die is rolled, there is a one-in-six chance for each number to be rolled.
3. Remind students about the Law of Large Numbers—the more times you conduct a probability experiment, the closer you are likely to get to the theoretical results.

Session 2: Off to the Races!

■ Getting Ready

1. Duplicate Double Dice Derby board with 12 horses for students, homework, and overhead.
2. Gather 12 markers for each pair of students and yourself.
3. Gather two dice, each a different color, for each pair of students and yourself.
4. Make chart to keep track of winners.

■ The Race Is ON!

1. Have they heard of Kentucky Derby? Today, they'll simulate a derby—the Double Dice Derby.
2. Put Double Dice Derby on overhead. Each of the 12 horses is competing to be the first to enter the winner's circle. Unlike a real derby, these horses move forward by the roll of two dice. The sum of the dice indicates which horse can move forward one space. Ask them to predict which horse they think will win. Demonstrate procedure and recording.
3. Distribute materials to partners and have them begin. Let students play until there is sufficient evidence that some horses are winning more frequently than others.
4. Focus attention on Results chart. Discuss, asking questions.
5. For homework, have students play game three more times with a family member or friend and record results of these races in journals.

Session 3: Keeping Track

■ Getting Ready

1. Duplicate one copy of Keeping Track chart for each student and make two overheads.
2. Select a pen in a different color from the one used by students for the initial race.
3. For Grades 4 and 5, pre-make number line from 0 to 1 with zero 36 inches from 1.

■ More Race Results

1. Have students share results from homework. Record on chart with different color pen. Ask for observations.
2. Build on responses to delve further into the reasons behind results.
3. Create a line graph of the race results.

■ Keeping Track

1. Put "blank" Keeping Track chart on overhead. It is a tool to show all possible combinations for the sums of two standard dice.
2. Point out numbers 1 through 6 along top horizontal line. These numbers represent the possible outcomes when one die is rolled. Then look along the vertical line on the left side of the chart. The outcomes for one die are also recorded. Have students help fill in first few numbers.
3. Distribute charts. As students work, have your completed Keeping Track overhead available, with colored pens.
4. Circulate and assist. Encourage students to look for patterns. When they complete chart, discuss results. How many combinations are possible when you roll two dice?
5. Connect the chart to the line graph. Ask how the two are related.
6. Provide a writing prompt.
7. For Grades 4 and 5, explore theoretical probability.

ACTIVITY 4: GAME STICKS

■ Getting Ready

1. Read over information on Native American Game Sticks.
2. Gather tongue depressors or craft sticks.
3. Create eight game sticks by drawing designs with colored markers on one side of each stick. Write your initials on other sides.
4. Play game a few times to learn it.
5. Gather counters for students and to model the game.
6. On chart paper, record number of counters or points for the outcomes.

■ Introducing Game Sticks

1. Say students will be learning a new game based on Native American games. It is called Game Sticks and uses four sticks decorated on only one side.
2. Show a few of the sticks you made.
3. For homework, students decorate a set of four sticks on one side. On the other side, they write their initials.

Session 1: Game Sticks in Action

1. Tell students they will learn how to play Game Sticks first as a whole class with your sticks.
2. Explain that the object is to obtain all 10 counters. Post chart with outcomes. Show and explain the outcomes of the sticks and connect to the related point(s).
3. Divide class into two teams and play game.
4. Be sure everyone understands the rules.

■ Playing the Game with a Partner

1. Have partners play game until one player has all the counters.
2. As students play, circulate and answer questions.
3. When most have played at least two times, focus for a class discussion.
4. Ask for their observations.
5. For homework, students play game at least two times with family member or friend.

Session 2: Investigating the Outcomes

1. Ask students for observations about the games they played for homework.
2. With student assistance, record all the possible outcomes they observed and draw those outcomes on the board.
3. With a partner, have students discuss which outcome they think is most likely and why.

■ Conducting an Experiment

1. Tell students they will do a probability experiment to help determine which outcome is most likely. They will toss sticks 15 times and a partner will record outcomes. Then they switch roles. Results for the class will be tabulated.
2. Model how to do this with a student.
3. Circulate as students conduct the experiment. Assist as needed. After both partners have tossed and tallied, hold a discussion.
4. Given the results of two experiments, ask what students think is the most common outcome.
5. Say that to have more data to consider, results from all their experiments will be collected. Have students combine their results.
6. Prepare chart to record numbers for each outcome.
7. After all results posted, calculate total number of times each outcome occurred.
8. What new insights do they have with this additional data? Can they say that a particular outcome occurs more frequently than others? Have partners discuss with one another.

■ Analyzing the Results

1. Analyze the possible outcomes as detailed in guide.
2. For third-grade students, end session with a writing prompt.

■ Delving Into the Theoretical (For Grades 4 and 5)

1. Have students total all possible ways for four two-sided sticks to fall.
2. Guide students in determining theoretical probability, expressed as fraction, for each outcome.
3. End session with writing prompt.

Assessment Suggestions

Selected Student Outcomes

1. Students are able to collect and organize data from probability experiments using tallies, charts, and graphs.

2. Students are able to determine appropriate scales for graphs to accurately display data and compare different representations of the same data.

3. Students increase their ability to justify conclusions and draw inferences based on data.

4. Students are able to describe events using the language of probability (likely, unlikely, always, never, equally likely, probable, and impossible).

5. Students can interpret data to make predictions about future events.

6. Students are able to generate the outcomes for simple probability experiments.

7. Students gain experience using a number from 0 to 1 to measure the likelihood of an event occurring.

8. Students improve in their ability to reason and communicate mathematically.

Built-in Assessment Activities

Penny Flip Journal Write: After students conduct an experiment, tally results, and interpret class data, they write their predictions for the total number of heads and tails and explain their reasoning for their predictions. Outcomes 1, 2, 5, 8

Penny Flip Graph (Homework): After students have analyzed the results of their Penny Flip experiment and created a class graph, students independently create a graph of the results. Outcomes 2, 8

Graph Again! Students make a second graph of the data from the Penny Flip experiment with attention to the scale. Students are encouraged to make a new type of graph, other than a bar graph. Outcomes 2, 8

Graphs from Home: Students look for graphs and analyze the graphs' scales. Then they interpret one graph and make statements and an inference about the data on it. Outcomes 2, 3, 8

Penny Flip Prediction: In their final journal writing for Activity 1, students predict the outcome of a penny toss experiment and explain their reasoning for their predictions. Outcomes 3, 5, 8

Track Meet: Students conduct a probability experiment using spinners and keep a record of winners from their experiments as well as record winners on a class graph. Outcomes 1, 5, 8

Fair and Unfair Spinners: For homework, students create a fair and an unfair spinner, applying what they have learned from the Track Meet spinners. Outcomes 4, 5, 8

Spinner Writing Prompts: Students respond to one of two prompts to articulate their understanding of the unfair spinner. Outcomes 4, 5, 6, 8

Fair or Not Fair?: Students write an independent response to the fairness of a spinner. After writing, students talk to classmates to share their thinking and then a class discussion is held. Outcomes 3, 4, 5, 6, 8

Graph of Roll ALL Six: As a class, students generate ideas for the scale of the class data collected for rolls of a die. After the graph is made, students discuss and analyze the data to make conclusions and inferences. Outcomes 1, 2, 3, 8

Double Dice Derby: In addition to conducting this experiment in class, students play the game three more times for homework, and bring in the results of the races. Outcomes 1, 4, 5

Wrapping Up Through Writing: Students give advice about which horse will win the Double Dice Derby. They are asked to explain using as much information as possible including graphs and charts. Outcomes 3, 4, 5, 6, 7, 8

Game Sticks at Home: Students play Game Sticks at least two times with a family member and collect data informally on the most and least common outcomes. Outcomes 1, 4, 5

Game Sticks Writing Prompt: Students are asked to determine the outcomes for a new version of Game Sticks and to assign a scoring system. Outcomes 3, 4, 5, 6, 7, 8

Two-Penny Toss (Going Further): Students play a game and analyze it through a probability lens to determine if it is a fair game. Outcomes 1, 3, 4, 5, 6, 8

Additional Assessment Idea

Mystery Bag: Pairs of students are given bags, each with a total of 10 cubes in two different colors. All the bags used by students have the same contents! Through a series of samplings, students will predict the number of cubes of each color in the bag. Start with a set of cubes that includes 3 of color A and 7 of color B. Have students conduct a replacement sampling with a partner. One person samples cubes and the other records the results. Student will reach in the bag, pull out a cube, and partner will record the color. Then put the cube back into the bag before sampling again! Shake the bag, sample again, and record result. Repeat until 10 samplings have been made. Based on their sampling, can they predict what is in the bag? Students switch roles and repeat the experiment. With the results of two experiments, they predict what the cube distribution is in the bag. Then as a class, gather all the results of the experiment. Again have students predict what is in the bag and explain why they chose that distribution.
Outcomes: 1, 3, 4, 8

Resources and Literature Connections

Materials

- Large Foam and Standard Dice
  ETA Cuisenaire
  www.etacuisenaire.com
  1-800-445-5985

- Standard Dice and Non-standard Dice
  Classroom Direct
  www.classroomdirect.com
  1-800-248-9171

Non-standard dice, with different markings or with more or less than six sides, are available from teacher supply and game stores.

Data-Related Web Sites

http://www.edhelper.com/graphs.htm
This site has a data sheet to support students in taking a survey of favorite snack items and then using the data to create a graph. The site also features additional graphing activities (bar graphs, converting tally chart into a graph, etc.) that are organized by grade levels.

http://www.usatoday.com
This paper features graphs in each of each four sections daily. At the bottom of the homepage, the **USA TODAY Snapshot** links to a selected daily graph and invites the user to respond to the question posed in the graph.

http://www.shodor.org/interactivate/activities/prob/index.html
This interactive site provides both graphing activities and probability experiments.

http://www.mathleague.com/help/data/data.htm
This site provides information on and examples of bar, circle, and line graphs.

http://www.bized.ac.uk/dataserv/penndata/graphs/index.htm
This site generates bar graphs; the one that would likely be of interest to students involves the population of a selected country from 1950 to 2000.

Related Curriculum Materials

Games of the Native American Indians
by Stewart Culin. Dover Publications, New York, 1975. IBSN: 0-486-23125-9.

How to Lie With Statistics
by Darrell Huff illustrated by Irving Geis. W. W. Norton and Company; Reissue edition, September 1993. ISBN: 0393310728

Investigations in Number, Data, and Space.
Scott Foresman, 2006. (Developed at TERC and funded by the National Science Foundation.)
Web: www.investigations.scottforesman.com
Phone: 1-800-552-2259

This curriculum includes books on data, graphing, and probability for Grades 3–5.
Grade 3: *From Paces to Feet* (Measuring and Data)
Grade 4: *Changes Over Time* (Graphs)
Grade 4: *The Shape of the Data* (Statistics)
Grade 4: *Three Out of Four Like Spaghetti* (Data and Fractions)
Grade 5: *Data: Kids, Cats, and Ads* (Data Analysis)
Grade 5: *Between Never and Always* (Probability)

Navigating Through Data Analysis and Probability in Grades 3-5 (includes CD-ROM).
National Council of Teachers of Mathematics (NCTM), 2002. IBSN: 0-87353-521-9

Principles and Standards for School Mathematics. NCTM, 2000. IBSN: 0-87353-480-8.

Probability Model Masters by Dale Seymour. Dale Seymour Publications, June 1991. ISBN: 0866515372. More than 100 models for modeling probability concepts.

Professional Standards for Teaching Mathematics. NCTM, 1991. IBSN: 0-87353-307-0.

What Are My Chances? Book A (Grades 4-6) by Albert P. Shulte and Stuart A. Choate. Creative Publications, Mountain View, California, 1991. ISBN: 0884880834

Fiction for Students

Alice

by Whoopi Goldberg; illustrated by John Rocco
Bantam Books, New York. 1992
Grades: 2–6

This book not only highlights its author's well-known comedic skills, it also conveys a lesson about friendship and some statistical wisdom relating to sweepstakes and their deceptive enticements. Alice enters "every sweepstakes, every giveaway, every contest," because she wants to be rich. She lives in New Jersey and one day is notified she has won a sweepstakes. Alice convinces her friends to go with her on an odyssey to New York City to collect the prize. What happens makes for a rollicking adventure, which your students will enjoy at the same time as they realize the probability and statistics lessons they are learning in class have lots of applications in the real world.

Anno's Three Little Pigs

by Mitsumasa Anno and Tuyosi Mori
The Bodley Head, London. 1985.
Grades: 4-8

Socrates, a wolf, attempts to catch one of three pigs for his wife's dinner. These three pigs collectively own five cottages. With the help of his frog-friend, the mathematician Pythagoras, Socrates tries to determine the possible cottages the pigs might be in. As the story unfolds, the illustrations show the many possible locations of the pigs and in doing so show, visually and clearly, the difference between permutations and combinations. This type of mathematics, known as combinatorial analysis, forms the basis for computer programming and problem-solving and is explained on a more advanced level in the back of the book.

Back in the Beforetime: Tales of the California Indians

retold by Jane L. Curry; illustrated by James Watts
Macmillan Publishing Co., New York. 1987
Grades: 2–6

A retelling of 22 legends about the creation of the world from a variety of California Indian tribes. In the myth "The Theft of Fire," the animal people spend an evening gambling with the people from the World's End. After the animal people lose all they have, Coyote, in a final bet, wagers the animal peoples' fire stones. The outcome of that bet is the basis for the mythic explanation of how the animal people got fire. The story ties in with Activity 4.

Betcha!

by Stuart J. Murphy; illustrated by S. D. Schindler
MathStart Series, Harper Collins Publishers, Inc., NY. 1997
Grades 3-5

Two friends estimate quantities from jelly beans in a jar to the value of all the toys in a store window. A great book to use as a springboard into estimation activities.

The Best Vacation Ever
by Stuart J. Murphy; illustrated by N. B. Westcott
MathStart Series, Harper Collins Publishers, Inc., NY. 1997
Grades 2-4

This book models how to conduct a survey and compile the results to determine the option most preferred by the group surveyed, which in this case is the location for the best vacation ever. Easy to read.

Cloudy With a Chance of Meatballs
by Judi Barrett; illustrated by Ron Barrett
Atheneum, New York. 1978
Grades: K–4

A hilarious look at weather conditions in the town of Chewandswallow, which needs no food stores because daily climatic conditions bring the inhabitants food and beverages, such as a storm of giant pancakes or an outpouring of maple syrup. This book presents a non-threatening way to look at predictions and to introduce the numeric scale for labeling theoretical probabilities. Students can follow up the story by listening to weather reports and charting the accuracy of meteorologists. They can also use The Cloud Book by Tomie dePaola to observe and chart clouds, one aspect of weather patterns.

Do You Wanna Bet? Your Chance to Find Out About Probability
by Jean Cushman; illustrated by Martha Weston
Clarion Books, NY. 1991
Grades 4–6

Two boys find that everyday events and activities involve chance and probability. The dialogue between them sparks the reader to predict and then find out what happens. In addition to experiments that connect to activities in the guide (penny flip, roll of dice), reading this book could spark other probability experiments.

The Hundred Penny Box
by Saron Bell Mathis; illustrated by Leo and Diane Dillon
Puffin Books, Viking Penguin, Inc, New York. 1975.
Grades: 3-6

Michael's love for his 100-year-old great-great-aunt, who lives with his family, leads him to intercede with his mother, who wants to throw out her old penny box and buy a new one. Each of the hundred pennies in Aunt Dew's box carries memories of one year of her life, and Michael listens attentively as Aunt Dew recounts her life stories through each one. Connects to the Penny Investigations in Activity 1.

Jumanji
by Chris Van Allsburg
Houghton Mifflin, Boston. 1981
Scholastic Books, New York. 1988
Grades: K–5

A bored brother and sister left on their own find a discarded board game (called Jumanji) that turns their home into an exotic jungle. A final roll of the dice for two sixes helps them escape from an erupting volcano. The story relates to the horse racing game in this guide—in both, a roll of the dice determines an important outcome.

People
by Peter Spier
Doubleday, New York. 1980
Grades: Preschool–6

Here's an exploration of the differences between (and similarities among) the billions of people on earth. It illustrates different noses, different clothes, different customs, different religions, different pets, and so on. This is a great book to use in collecting statistics and creating graphs about characteristics of people. Pairs of students can investigate the occurrence in their class of a physical feature (hair type, eye color, etc.), preference (types of food), or other distinguishing attribute (where one lives), and report their findings to the class.

REVIEWERS *(of first edition)*

We warmly thank the following educators, who reviewed, tested, or coordinated the trial tests in manuscript or draft form. Their critical comments and recommendations, based on classroom presentation of these activities nationwide, contributed significantly to this GEMS publication. (The participation of these educators in the review process does not necessarily imply endorsement of the GEMS program or responsibility for statements or views expressed.) Classroom testing is a recognized and invaluable hallmark of GEMS curriculum development; feedback is carefully recorded and integrated as appropriate into the publications. WE THANK THEM ALL! ■

Note: *Teachers who field tested the 2006 revised version are listed on page iii.*

ALASKA
Coordinator: Cynthia Dolmas Curran

Iditarod Elementary School, Wasilla
Cynthia Dolmas Curran
Jana DePriest
Christina M. Jencks
Abby Kellner-Rode
Beverly McPeek

Sherrod Elementary School, Palmer
Michael Curran
R. Geoffrey Shank
Tom Hermon

CALIFORNIA
GEMS Center, Huntington Beach
Coordinator: **Susan Spoeneman**

College View School, Huntington Beach
Kathy O'Steen
Robin L. Rouse
Karen Sandors
Lisa McCarthy

John Eader School, Huntington Beach
Jim Atteberry
Ardis Bucy
Virginia Ellenson

Issac Sowers Middle School, Huntington Beach
James E. Martin

San Francisco Bay Area
Coordinator: Cynthia Eaton

Bancroft Middle School, San Leandro
Catherine Heck
Barbara Kingsley
Michael Mandel
Stephen Rutherford

Edward M. Downer Elementary School, San Pablo
M. Antonieta Franco
Nancy Hirota
Barbara Kelly
Linda Searls
Emily Teale Vogler

Malcolm X Intermediate School, Berkeley
Candyce Cannon
Carole Chin
Denise B. Lebel
Rudolph Graham
DeEtte LaRue
Mahalia Ryba

Marie A. Murphy School, Richmond
Sally Freese
Dallas Karahalios
Susan Jane Kirsch
Sandra A. Petzoldt
Versa White

Marin Elementary School, Albany
Juline Aguilar
Chris Bowen
Lois B. Breault
Nancy Davidson
Sarah Del Grande
Marlene Keret
Juanita Rynerson
Maggie J. Shepherd
Sonia Zulpo

Markham Elementary School, Oakland
Alvin Bettis
Eleanor Feuille
Sharon Kerr
Steven L. Norton
Patricia Harris Nunley
Kirsten Pihlaja
Ruth Quezada

Martin Luther King, Jr. Jr. High
-Sierra School, El Cerrito
Laurie Chandler
Gary DeMoss
Tanya Grove
Roselyn Max
Norman Nemzer
Martha Salzman
Diane Simoneau
Marcia Williams

Sleepy Hollow Elementary School, Orinda
Lou Caputo
Marlene Fraser
Carolyn High
Janet Howard
Nancy Medbery
Kathy Mico-Smith
Anne H. Morton
Mary Welte

Walnut Heights Elementary School, Walnut Creek
Christl Blumenthal
Nora Davidson
Linda Ghysels
Julie A. Ginocchio
Sally J. Holcombe
Thomas F. MacLean
Elizabeth O'Brien
Gail F. Puleo

Willard Junior High School, Berkeley
Vana James
Linda Taylor-White
Katherine Evans

GEORGIA
Coordinator: Yonnie Carol Pope

Dodgen Middle School, Marietta
Linda W. Curtis
Joan B. Jackson
Marilyn Pope
Wanda Richardson

Mountain View Elementary School, Marietta
Cathy Howell
Diane Pine Miller
Janie E. Stokes
Elaine S. Toney

NEW YORK
Coordinator: Stanley J. Wegrzynowski

Dr. Charles R. Drew Science Magnet, Buffalo
Mary Jean Syrek
Renée C. Johnson
Ruth Kresser
Jane Wenner Metzger
Sharon Pikul

Lorraine Academy, Buffalo
Francine R. LoGrippo
Clintonia R. Graves
Albert Gurgol
Nancy B. Kryszczuk
Laura P. Parks

OREGON
Coordinator: Anne Kennedy

Myers Elementary School, Salem
Cheryl A. Ward
Carol Nivens
Kent C. Norris
Tami Socolofsky

Terrebonne Elementary School, Terrebonne
Francy Stillwell
Elizabeth M. Naidis
Carol Selle
Julie Wellette

Wallowa Elementary School, Wallowa
Sherry Carman
Jennifer K. Isley
Neil A. Miller
Warren J. Wilson

PENNSYVANIA
Coordinator: Greg Calvetti

Aliquippa Elementary School, Aliquippa
Karen Levitt
Lorraine McKinin
Ted Zeljak

Duquesne Elementary School, Duquesne
Tim Kamauf
Mike Vranesivic
Elizabeth A. West

Gateway Upper Elementary School, Monroeville
Paul A. Bigos
Reed Douglas Hankinson
Barbara B. Messina
Barbara Platz
William Wilshire

Ramsey Elementary School, Monroeville
Faye Ward

WASHINGTON
Coordinator: David Kennedy

Blue Ridge Elementary School, Walla Walla
Peggy Harris Willcuts

Prospect Point School, Walla Walla
Alice R. MacDonald
Nancy Ann McCorkle